T0210782

Greenland's Economy and Labour Markets

This book explores structural changes in Greenland's economy and labour markets due to the transformative effects of climatic changes and growing international attention. It offers multidisciplinary perspectives from economists, sociologists, and political scientists to demonstrate how the Greenlandic economy works.

Due to an increasing focus on the Arctic area and Greenland in particular, the book seeks to understand the functioning and dynamics of Greenland's labour economy, as well as the challenges that arise from the melting ice and internationalisation. It fills a substantive gap in the existing literature by compiling research on these critical subjects and exploring current and future opportunities for labourers. Today, Greenland is reliant on large financial subsidies from Denmark to provide for a large share of its national budget. This fuels Greenland's political ambition to gain greater independence from Denmark, which requires more private sector growth to develop a sustainable economy. This book thus contains an exhaustive introduction to important business development themes such as macroeconomics, markets, labour supply, labour market policies, and institutions and considers Greenland's colonial past, great Inuit heritage, and unique geography and nature to re-shape its economy and labour markets. Informed by a lucid writing style, each chapter casts light on different economic and social issues of Greenland.

This is the first international book on Greenland's economy which discusses its geopolitical importance and prospects for the Arctic region. It will be a valuable point of reference for students and academics of economics, Arctic research and political economy.

Laust Høgedahl is an Associate Professor at the Centre for Labour Market Research (CARMA), Department for Politics & Society, Aalborg University, Denmark. He holds a PhD degree in Political Science and does research within the academic fields of industrial relations and labour market regulations.

Routledge Research in Polar Regions

Series Editor: Timothy Heleniak, *Nordregio International Research Centre, Sweden*

The Routledge series in Polar Regions seeks to include research and policy debates about trends and events taking place in two important world regions: the Arctic and Antarctic. Previously neglected periphery regions, with climate change, resource development and shifting geopolitics, these regions are becoming increasingly crucial to happenings outside these regions. At the same time, the economies, societies and natural environments of the Arctic are undergoing rapid change. This series seeks to draw upon fieldwork, satellite observations, archival studies and other research methods which inform about crucial developments in the Polar regions. It is interdisciplinary, drawing on the work from the social sciences and humanities, bringing together cutting-edge research in the Polar regions with the policy implications.

Resources, Social and Cultural Sustainabilities in the Arctic
Edited by Monica Tennberg, Hanna Lempinen, and Susanna Pirnes

Arctic Sustainability, Key Methodologies and Knowledge Domains
A Synthesis of Knowledge I
Edited by Jessica K. Graybill and Andrey N. Petrov

Food Security in the High North
Contemporary Challenges Across the Circumpolar Region
Edited by Kamrul Hossain, Lena Maria Nilsson, and Thora Martina Herrmann

Collaborative Research Methods in the Arctic
Experiences from Greenland
Edited by Anne Merrild Hansen and Carina Ren

Greenland's Economy and Labour Markets
Edited by Laust Høgedahl

For more information about this series, please visit: www.routledge.com/ Routledge-Research-in-Polar-Regions/book-series/RRPS

Greenland's Economy and Labour Markets

Edited by
Laust Høgedahl

Routledge
Taylor & Francis Group

LONDON AND NEW YORK

First published 2022
by Routledge
2 Park Square, Milton Park, Abingdon, Oxon OX14 4RN

and by Routledge
605 Third Avenue, New York, NY 10158

Routledge is an imprint of the Taylor & Francis Group, an informa business

British Library Cataloguing-in-Publication Data
A catalogue *record* for this book is available from the British Library

Library of Congress Cataloging-in-Publication Data
A catalog record has been requested for this book

ISBN: 978-0-367-51619-2 (hbk)
ISBN: 978-0-367-51627-7 (pbk)
ISBN: 978-1-003-05463-4 (ebk)

Typeset in Times New Roman
by KnowledgeWorks Global Ltd.

Contents

List of Figures

List of Tables

Contributors

Torben M. Andersen is a professor at the Department of Economics, Aarhus University, Denmark. He holds a PhD from Université Catholique de Louvain, Belgium (1986). He is currently the chair of the Economic Council of Greenland appointed by the Government of Greenland.

Javier L. Arnaut is an assistant professor at the University of Greenland (Ilisimatusarfik), Nuuk, Greenland. His research interests include economic history, environmental economics, political economy, and gender studies. He received a PhD in Economics, from the University of Groningen, The Netherlands.

Laust Høgedahl is an associate professor at the Centre for Labour Market Research (CARMA), Department for Politics & Society, Aalborg University, Denmark. He holds a PhD degree in Political Science and does research within the academic fields of industrial relations and labour market regulations.

Helle Holt (PhD) is a senior researcher at VIVE – The Danish Center for Social Science Research. Research topics: Working life studies, qualitative evaluation of different measurements for the unemployed vulnerable people in the Danish and the Greenlandic employment systems.

Cecilie Krogh is a PhD fellow, Aalborg University. Cecilie is a sociologist investigating the labour market, focusing on employer attitudes and behaviours in recruitment processes.

Rasmus Lind Ravn is a postdoctoral researcher at The Department of Politics and Society at Aalborg University. His main areas of research are disadvantaged unemployed and active labour market policies.

Niels Jørgen Mau Pedersen is a project director VIVE – The Danish Center for Social Science Research. He has previously been the Head of Department on Local Government Economics at the Ministry of Welfare.

Helene Pristed Nielsen is an associate professor at FREIA Centre for Gender Research, Aalborg University, Denmark. Her research interests revolve around gender, mobility and place. Previous publications include the edited volume "Gender and Island Communities" (with Firouz Gaini) published by Routledge in 2020, as well as the report "Equality in Isolated Labour Markets" (with Erika Anne Hayfield and Steven Arnfjord) published by TemaNord 2020.

Frederik Thuesen is a senior researcher at VIVE – The Danish Center for Social Science Research specialised in evaluating and analysing the impact of labour market policies, employment, and working-life issues.

Foreword

By Mogens Lykketoft

This intriguing book about Greenland's economy and labour market elicits all the challenges the 60,000 people face on their road towards increased independence. The people of Greenland are scattered across 2 million km² of untamed nature with an icecap that is rapidly melting because of climate change.

The challenges in making the economy sustainable are great due to vast distances and dramatic climate changes. Greenland's primary industry is fishing. Post COVID-19, when affluent people are able to travel again on a large scale, tourism can become a significant source of income for the country. However, tourism and mining to extract minerals and possible energy reserves require a substantial investment in infrastructure. A prerequisite for business and trade development is for Greenlanders to obtain and improve vocational skills; otherwise, the small population must accept a far greater immigration rate of foreign labourers.

The climate changes and rivalry between superpowers to access resources and military presence in the Arctic regions raise new questions – What is the Denmark's role in Greenland? How will Greenland and Denmark balance Danish interference? Whether to stay in the Danish Realm is also debated. The United States' role as the fundamental military power in Greenland has been constant since 1941; however, we have witnessed the United States' desire to gain a greater influence in Greenland, including Greenlandic civil society.

The Danish Realm is, for those of us who know Greenland, an immense source of fortune and pleasure. The people and their interaction with the magnificent arctic nature are unique and fascinating. We hope that our bond, which prevails despite new challenges, can be preserved for future generations.

Very few people who live their entire lives in Greenland have solely Inuit genetic backgrounds, after 300 years of coexistence in the Kingdom of Denmark. Both origin and history connect us.

Denmark has invested significantly in improving living conditions in Greenland, and most Danes are not reluctant to continue providing considerable economic assistance to Greenland.

Throughout the past seven decades, living conditions in Greenland have increased considerably. The improvements have not necessarily been on Greenlanders' own terms, and many people have had to give up life as hunters to become individuals in modern societies – this has also had a negative impact. Inevitably, most Greenlanders gather in few urban communities; at least one-third of the population will soon be living in the capital, Nuuk.

A strong sense of a unique Greenlandic identity understandably fosters a desire to become more self-governing.

Home Rule was therefore implemented in 1979 and considerably extended with the Self-Government Act in 2009. Greenland is able to undertake all national affairs, except the Monarchy, the Supreme Court, foreign affairs, and defence policy. Furthermore, nowadays, Greenlandic representatives always participate when foreign affairs concerning Greenland are negotiated with other countries.

Many people in Greenland regard complete independence as the long-term goal.

If Greenland demands complete independence in the future, then of course, Denmark would accept this, despite what the Danish Constitution says on this matter. The Greenlanders were not included when the Constitution of Denmark was written!

Do we not have mutual interests that are more valuable than those to be exerted by superpowers towards a self-governing Greenland? It is also difficult to imagine Greenland replacing its head of state, considering the very strong, warm feelings that exist between the Greenlanders, and the Danish Crown Prince Frederik.

Preface

Climate change, abundant resources, and geopolitical importance have brought Greenland into the limelight on the world stage in recent years. In August 2019, US President Donald Trump showed serious interest in purchasing Greenland from Denmark, sparking a diplomatic spat between Denmark and the United States. Tension grew when the United States launched a new consulate in Nuuk and offered Greenland economic relief during the COVID-19 pandemic. Greenland is in a highly transformative period due to climate changes and growing international attention, and these changes will arguably have a big impact on Greenland's economy and labour markets in future. This volume explores Greenland's economy and labour markets from a variety of disciplinary perspectives, including economics, sociology, and political science. Our goal is twofold. First, we want to illustrate how the Greenlandic economy and labour markets work. We find that this type of work is missing in existing literature; thus, research on these important matters is in demand due to the growing focus on the Artic area and Greenland in particular. Second, we want to explore the opportunities and challenges that arise from the melting ice and internationalisation. These questions are highly interlinked with the growing political ambition in Greenland to become more independent from Denmark. Today, Greenland is reliant on large financial subsidies from Denmark to make up a large share of its national budget. More independence from Denmark requires more private-sector growth to develop a sustainable economy; hence, new areas of business are being pursued.

This book contains an exhaustive introduction to important areas such as macroeconomics, markets, labour supply, labour market policies, and institutions. We show how Greenland's colonial past, great Inuit heritage, and unique geography and nature have shaped the economy and labour markets. Each chapter sheds lights on different elements and will be of value to those with an interest in economic and social issues in Greenland.

1 Introduction

The peculiarities of Greenland's economy and labour markets and the driving forces of change

Laust Høgedahl

Greenland is a self-governing part of the Kingdom of Denmark, and while its government decides on most domestic matters, foreign and security policies are handled by the Danish Government in Copenhagen. However, a World War II defence treaty between Denmark and the United States gives the US military virtually unlimited rights in Greenland at America's northernmost base, Thule Air Base. Yet, global warming and contracting ice have opened up access to the Northwest Passage for commercial shipping, exposed new potential offshore oil mining sites in the Artic Sea, and eased the logistics of accessing precious minerals. Hence, not only American but also Russian and especially Chinese companies have tried to gain economic footholds in Greenland and the surrounding seas (Sejersen, 2015).

The growing political and economic interests in Greenland have also sparked a strong domestic desire for more independence from Denmark. As an old colony, Greenland relies on 3.9 billion DKK in annual subsidies from Denmark, which make up about 55% of its annual budget. A stronger degree of independence is closely tied to a more self-sustaining economy based on private sector growth. However, Greenland is a vast country, spanning 2,166,086 km² with a small population of only 55,000 inhabitants mostly concentrated along the west coast. The capital, Nuuk, is the largest city, holding around 18,300 people. The geography and the size of the country make mobility and transportation challenging.

The economic development has been very positive in recent years; however, public and especially private employers in Greenland are experiencing a labour shortage in spite of a rather high unemployment level of around 6%; a relatively large share of the population is not part of the workforce. The private economy is also heavily dependent on fishing and its downstream industries, making the Greenlandic economy vulnerable to fluctuation in fish prices and stock. Therefore, new areas of business are being pursued, such as minerals, adventure tourism, renewable energy, and pure water exports, combined with policy reforms aimed at increasing the labour supply and easing access to foreign labour (Government of Greenland, 2020).

In this introductory chapter, we present the peculiarities of Greenland's economy and labour markets. In many ways, conventional theories of economics, labour market regulations, and working life fit poorly with the Greenlandic reality. This is due to a mix of structural, institutional, cultural, and political circumstances that have shaped Greenland into a unique case over the course of time, sharing traits with both Nordic countries (especially Denmark) and Inuit communities in North America.

First, this opening chapter introduces a number of key concepts and terms relevant to studying any labour market, with a focus on Greenlandic conditions and peculiarities: *structural* and *institutional conditions*, *demand*, and *supply of labour*. We explore these key concepts in the following chapters. Second, this chapter involves a number of structural and institutional "driving forces of change" that will continue to shape the Greenlandic economy and labour market substantially. In particular, we address the consequences of climate changes and internationalisation. It is important to stress that these driving forces of change are not all deterministic but rather create new possibilities and challenges for the Greenlandic economy and labour market. Primarily, change follows when policymakers regulate or de-regulate.

This introductory chapter ends with a brief overview of the following chapters and by summarising the most important overall conclusions.

1.1 Structural and institutional conditions for Greenland's economy and labour markets

The *structural conditions* of any country are obviously important in terms of its labour market and overall economy. In economic and political economic theory, structural conditions include a variety of predetermined factors given to any national economy (Stiglitz, 2010). Some countries benefit from vast pools of raw material – some renewable, such as fish, and others limited, such as oil, gas, or minerals. Other countries are landlocked, limiting their means of importing and exporting goods. Still other countries, such as Greenland, are not connected with the mainland. Hence, structural conditions are important in terms of shaping industries and businesses for any national economy (e.g., shaping employment in the primary, secondary, and tertiary sectors). Structural conditions are not static but continuously transform, bringing *structural change* to an economy and related labour markets.

The structural conditions framing Greenland's economy and labour market are unique. First, the sheer size of the country in relation to the size of the population is stunning. Greenland spans 2,166,086 km² – more than the states of Alaska and California combined. The country ranges from Cap Morris Jesup in the north, just 740 km from the North Pole, to Cap Farewell in the south, situated on the same latitude as the capital of Norway, Oslo. The widest point from west to east is 1,050 km, and more than 410,000 km² (around 19%) is covered by inland ice. Greenland is situated northeast of Canada and is therefore, geographically, a part of North America in spite of its historical, political, and economic ties to Denmark and the rest of Europe.

Greenland holds 55,000 inhabitants, mostly concentrated along the west coast. The largest city and capital, Nuuk, holds around 18,300 people. The vast distances and impassable geography of mountain ranges, fjords, and ice make transportation challenging, especially considering the changeable weather. Even though Greenland technically is an island (the world's largest), it is in fact more precise to describe the country as an archipelago due to the logistic challenges posed by the geography (see Chapters 6 and 7). Labour mobility is a challenge for most countries; however, the Greenlandic labour market is extremely segmented in this regard. It is therefore more precise to describe the Greenlandic labour market as multiple – and very isolated – labour markets with extremely different structural conditions. The seasons in Greenland also strongly affect the economy and labour market, as fishing and other activities, such as construction, are dependent on the weather.

The country's diverse geography also shapes the nature of various industries. The unique geography and geology of Greenland produce a number of natural resources in terms of fish (including shellfish, mainly prawns), oil, gas, and minerals. Fishing as a renewable, natural resource takes place throughout the country, though the west coast is home to most of the inshore commercial fishing. Farming and agriculture have long been present in southern Greenland, producing both meat (mainly sheep) and vegetables. In the north, the sun does not rise from late October to mid-February, setting limitations for outdoor activities. There are two active mines in Greenland – one ruby mine in Qeqertarsuatsiaat and a newly opened mine at Kangerlussuaq that produces anorthosite. Commercial extraction of minerals is still rather limited, yet new mining sites are seen as important areas of income and employment for the future, as the Danish granted competence in this policy area to the Greenlandic government in 2010.

Oil and gas are also found underground in the seas surrounding Greenland (Skjervedal, 2018). Hence, the Greenlandic government has actively been pursuing an offensive strategy to attract foreign investments in commercial extraction by offering tax cuts to so-called "first movers" willing to invest in developing new oilfields (Government of Greenland, 2020).

The public sector is the biggest sector and main employer in Greenland, employing around 40% of the workforce. Due to Greenland's colonial history, many companies remain state-owned. Fishing is the dominant industry besides public employment, with the state-owned Royal Greenland as the biggest employer in Greenland.

In terms of institutions, Greenland gained rights as an independent territory (amt) of the Kingdom of Denmark in 1953 and therefore not just a colony. A number of incremental reforms have since given Greenland more self-governance from Denmark, albeit foreign and security policy are still handled by the Danish Government in Copenhagen. In terms of economy, Greenland is dependent on annual subsidies from Denmark to make up half of its annual budget.

Regarding labour market regulation and institutions, Greenland can be seen as part of the Nordic model of industrial relations (Høgedahl, 2019). Wages are primarily negotiated by strong labour market (social) partners. On the employers' side, Grønlands Erhverv (GE) represents small and large companies within all sectors (apart from the public). On the employees' side, a number of trade unions are present in the labour market, with SIK as the biggest organisation of unskilled and skilled blue-collar workers. The labour market partners conclude collective agreements that regulate wages and other working conditions in all sectors. Strikes and lockouts are only allowed when collective agreements are concluded or renegotiated. When a collective agreement is in force, labour disputes must be solved by means of negotiation or mediation. This industrial "peace obligation" is also known in the rest of the Nordic countries. However, Greenland has a national statutory minimum wage set by national law, which is not the case in other Nordic countries. In Greenland, no wage coordination mechanism exists to ensure that the exporting industry dictates wage developments. Another important difference between the Nordic countries and Greenland is that no independent, private labour court exists in Greenland. Here, labour disputes are solved in the public court system or by private arbitration. It is also important to note that foreign labour is also regulated through law (see Chapter 5).

1.2 Supply and demand for labour

Any industrialised labour market has a demand and a supply side. The demand side comes from employers (public and private) that need labour to produce goods and services. The supply side is the labour available. The demand and supply for labour are also obviously pivotal in terms of unemployment and wage levels, although affected by labour market policies as well as employment and industrial relation systems.

Traditionally and historically, most Greenlandic families have been self-sufficient (self-employed) by fishing, hunting (including sealing and whaling), and trading. After World War II, the Greenlandic economy slowly transformed to a production-based economy, with a supply and demand for labour as a consequence (Thorleifsen, 2005). The first workers were predominantly employed at KGH Royal Greenland Trading (Den Kongelige Grønlandske Handel), a Danish state-owned company trading provisions and goods from settlements around Greenland (Marquardt, 2005). Since the 1950s, the Greenlandic labour supply has grown exponentially. The rather late industrialisation of the Greenlandic economy, compared to other western European countries, came with new challenges in terms of matching the supply and demand of labour. Education and labour market policies were needed to supplement the industrial policy (Thorleifsen, 2005), and the need for education policies is still very present today (see Chapter 8). The structural condition described above is also influencing education and training, as many young Greenlanders have to leave home at around 15 years old because their higher education takes place in another city; many small towns and settlements are often too small to hold larger schools. In 1953, Greenland adopted an educational system similar to the Danish system. Changes during recent decades within the country's primary sectors – fishing and hunting – mean that necessary changes have to take place to compete on a global level. Urbanisation (and thus, housing) is another specific example of Greenland's difficulty in keeping up with demand. Education is seen as a means to achieve better welfare, improvement of health, and global (political) power. However, many young Greenlanders must go abroad (for the most part, to Denmark) to get an education. Many do not return, creating a "brain drain" in terms of many much-needed professions (see Chapter 2).

In recent years, fish prices have been very favourable, creating a strong economic upturn in Greenland. This upturn has created more activities in other sectors, such as construction and private services, creating an even faster-growing demand for more labour. Greenlandic employers have therefore requested easier access to foreign workers, especially from Asia (China, Thailand, and the Philippines). The high demand for labour is also evident when looking at unemployment levels. Formally educated workers are in high demand, and even unskilled workers are very much needed, especially in the fish processing industry. Yet, a number of people in Greenland are still not part of the workforce due to a number of complicated issues (see Chapters 5, 7, and 9). These can be identified as social and psychological issues in combination with a lack of skills and education. However, research also indicates that Greenlandic employers, in many cases, wish to bypass native workers in favour of foreign workers (see Chapter 4).

Due to Greenland's history with Denmark as a colonial power, a close tie to the Danish labour market still exists today. Many workers with managerial responsibilities and higher education are native Danes working in Greenland (Government of Greenland, 2018). In this sense, the labour market is not only fragmented and segmented due to challenging geography, as discussed above, but also in terms of the need for foreign labour creating ties to Danish, Nordic, European, and Asian labour markets in various ways.

Another important element in terms of supply and demand for labour is the notion of work. In the world of work, the term is pretty well understood as an action by which a worker sells their labour to an employer for a given price to obtain a certain living standard. Other themes are, of course, relevant in relation to work, including a sociological emphasis on work as a way of forging identity. Yet, as Poppel (2005) notes, work as an activity or concept is different in an arctic setting. Due to cultural traditions, work as an activity has historically been seen in Greenland as something for those unable to provide for themselves in a hunter-gatherer society; since women were not considered hunters, they were compelled to work (see Chapter 8). Hence, some studies have found that men are very hesitant (even reluctant) to engage in certain types of paid work (e.g., the fish-process industry), seeing it as women's work (Poppel, 2005). When the notion of work is different from the general understanding, the concept of being unemployed becomes equally different, as Hansen and Tejsner (2016) showed in a study of northwest Greenland. We point out that being unemployed in a Greenlandic context is not the same as not working; one may still be providing for oneself by hunting or fishing, even if not partaking in formal paid work. The modern notion of working or not working in terms of paid work is therefore closely related to Inuit culture. These notions of work are important to bear in mind to understand labour mobility patterns, such as job turnover and unemployment levels, in Greenland.

1.3 The driving factors of change – climate and internationalisation

Greenland is at the absolute forefront in terms of climate change and internationalisation. In many ways, these two concepts are very much linked and correlated in Greenland (Nuttall, 2017).

Climate changes are obvious in many parts of Greenland, and the inland ice bears witness to many years of climate change, as revealed in ice cores extracted by climate researchers (Jansen et al., 2016). Contracting and melting ice means rising water levels and new weather patterns all around the world. However, climate changes are also affecting the economy and labour market in many ways locally in Greenland (Nuttall, 2010). Changes offer challenges as well as opportunities, as precisely formulated by politician Josef Motzfeldt:

> Climate change has already opened new areas for the exploitation of mineral resources as the ice cap is retreating. And in combination with the political and economical control of our mineral resources it will open new opportunities for Greenland to gain more economical and political independence from Denmark.... We have to choose on the one hand between unrestricted exploitation of our resources in order to gain more independence, and on the other hand the protection of our nature, which is so dear to us in order to maintain our cultural heritage in the shape of a close interrelationship between human activity and changes in the environment.
>
> (Josef Motzfeldt, as cited in Nuttall, 2008, p. 70)

This quote illustrates the built-in paradox presented by climate change. Contracting ice offers new economic opportunities on land and at sea, but with possible consequences for the environment. Climate change has already opened access to the Northwest Passage for commercial shipping, exposed new potential offshore oil mining sites in the Artic Sea, and eased the logistic access to precious minerals. New areas of business are being pursued, closely linked to the ambition of more private-sector growth enabling more economic independence from Denmark. However, in terms of the labour market, climate changes are already affecting Greenland. More unpredictable weather patterns make logistics and transportation by air and sea more challenging. Contracting ice during winter can also hamper mobility, since travelling by ice on snowmobiles is limited and loading/load-out of passengers and goods on sea ice near settlements becomes riskier. However, changing seasons and warmer temperatures are also creating new opportunities for farming in southern Greenland, creating new jobs and economic activities.

Greenland's ambition in terms of political and economic independence from Denmark through more private-sector growth in the wake of climate change is also attracting attention from foreign investors. The global attention on Greenland is underpinned by foreign governments' geopolitical interests in the Arctic. Chinese companies have shown great interest in mining and large-scale construction projects, including new airports. The US has opened a consulate in Nuuk and is working to increase the American presence in Greenland. Currently, three new international airports are being built in Greenland, and this massive investment in logistics is arguably going to have a great accelerating effect on the internationalisation of Greenland's economy and labour markets in the near future.

1.4 The chapters

This volume contains eight chapters concerning various elements of Greenland's economy and labour markets. Chapter 2 deals with macroeconomics of Greenland and present prospects and challenges for the overall economy. Readers will get a thorough introduction to the fundamentals of Greenland's macroeconomics, including an overview of the structure of the Greenlandic economy, its recent economic performance, and future prospects. We also point to some pivotal challenges in terms of increasing inequality in the country as well as the lack of education and vocational skills among the population. We conclude that improved education will contribute to a lessening of social problems and improve labour market prospects. If combined with the expansion of the private sector, this may also contribute to better public finances via increased tax payments and lower social expenditures. Thus, there is room for action to put the economy on a path that eventually may lead to a self-sustaining economy. Chapter 3 shows that the current economic situation of Greenland cannot be properly understood without an adequate understanding of its historical background. By applying a political economic approach, we argue that Greenland's economic development is deeply rooted in the evolution of economic relations originating three centuries ago; some of the forces that shaped those relations are still present. The existing organisational structure that governs the economy is essentially path-dependent on

Danish-Greenlandic historical relations, as the chapter shows. We conclude that an imminent transformation in Greenland's political economy is unfolding in alignment with the desire for economic independence. However, transformation is an endogenous factor that is encouraged by exogenous factors such as the natural-resource transition, which is heightened by the effects of climate change and the rise in global demand and supply shocks of mineral commodities. Successfully harnessing the benefits of the new transition may rely on the ability to transform the country's economic institutions to a type that promotes a more diversified and environmentally sustainable structure. In Chapter 4, we analyse the Greenlandic employment system by looking at Majoriaq, a frontline labour market institution with the thorny task of facilitating good matches among unemployed workers and labour-seeking employers. The chapter shows that a challenging matrix of demography and geography – few people dispersed on the lengthy fringes of an enormous ice cap – characterises the labour market in Greenland. Nuuk, along with Sisimiut and Ilulissat, is the centre of economic growth, while many smaller towns struggle with stagnating local economies and high unemployment rates. Aggravating the challenge, the working-age population in Greenland has a relatively low level of education, while an increasing number of jobs require increasingly advanced skills. We conclude that employees in the Majoriaq need more support, as their skills are insufficient in relation to the challenges they encounter. In Chapter 5, we look at Greenlandic labour supply by applying new, unique survey data and conclude that Greenland has a relatively low unemployment rate when applying the Labour Force Survey approach. This is mainly because many Greenlanders in the working population can be considered inactive, meaning that they are not actively searching for jobs or are unable to take up paid work within 14 days. The latter is important due to a number of reasons, including health and social issues. The chapter also shows that Greenlanders do not stay a long time in the same job compared to other countries, calling for better retention strategies. Greenlanders also tend to work long hours, and few people work part time. Many are self-employed – mainly men fishing from small open boats – however, the chapter shows that many women desire to be self-employed, which might be a key to more private-sector growth in the future. In Chapter 6, we look at the social security system as a whole and point to important structural and institutional barriers to labour market participation. We describe a broad range of barriers to labour market participation, including aspects of the equal treatment of citizens across local authorities, administrative procedures, matters of transparency for potential recipients of public assistance, interaction problems between tax regulation and income transfer schemes, and the organisational framework surrounding the social security system. Chapter 7 also contains a description of barriers to labour market participation, but from an individual perspective. We find that in spite of the peculiarities of the Greenlandic labour markets, the "usual suspects" are indicated when it comes to individual factors associated with barriers to labour market participation. Unskilled labourers living in rural areas are unlikely to participate in the labour market, and we find a lower employment rate of women (compared to men), even taking educational attainment, age, and geography into account. However, gender differences evened out when controlling for participation in education. The lower employment rate for women is thus due to the fact that Greenlandic women participate in education to a greater extent than Greenlandic men. In Chapter 8, we explore gender-related issues connected to Greenland's labour market more in depth. We reach a number of interesting conclusions, including a strong vertical gender segregation (between high- and low-paid jobs) in Greenland. This segregation is not only gendered but also partly ethnically based, with more ethnic Danes at the top rungs of public and private management. Finally, in Chapter 9, we explore an important and well-debated issue in terms of so-called NEETs (not in employment nor education or training) among young people. The chapter shows that even a few years as a NEET can have long-lasting "scare effects" in terms of unemployment later in life. The chapter also contains a description of how the current employment system is tackling this problem, and we point out some areas that need attention from policymakers and other stakeholders.

1.5 Conclusions

A number of conclusions can be drawn by considering the chapters in this volume. It is clear that Greenland is experiencing a very transformative period due to increasing international attention due to the geopolitical importance of Greenland, which has been amplified by contracting ice following climate changes. A strong domestic political ambition of more independence from Denmark is linked to more private-sector growth, which is not an easy task. We show from various perspectives that the educational level of the working population is crucial in many ways. An increase in educational and vocational levels will have a number of positive effects on the quality of work, equality, and the overall economy. However, there are no quick or easy solutions to this thorny task; it will arguably take time and large investments. Apart from educational levels, it is also clear that a number of other barriers are at play in terms of connecting supply and demand of labour in Greenland. A relatively large share of the working population is not part of the labour market in spite of a booming economy and a high demand for any type of labour in most parts of the country. This volume lists a number of reasons for this. Some might be linked to motivational effects created by the relation between the social security and tax systems. However, it is also evident that a number of health and social issues stand between many Greenlanders and labour market participation. This is also clear when looking at the employment system Majoriaq is facing in terms of the challenges of promoting employment. It is a challenge to balance a growing demand by employers for more easy access to foreign labour while a relative large part of the domestic working population can be considered inactive.

References

Government of Greenland. (2018). *Labour Market Report 2016–2017.* https:// naalakkersuisut.gl/~/media/Nanoq/Files/Publications/Arbejdsmarked/DK/ Arbejdsmarkedsredeg%C3%B8relse%202016-17%20DK.pdf

Government of Greenland. (2020). *Labour Market Report 2018–2019.* https:// naalakkersuisut.gl/~/media/Nanoq/Files/Publications/Arbejdsmarked/DK/ Arbejdsmarkedsredeg%c3%b8relsen%202018-2019%20DK.pdf

Hansen, A. M., & Tejsner, P. (2016). Challenges and opportunities for residents in the Upernavik District while oil companies are making a first entrance in Baffin Bay. *Arctic Anthropology, 53* (1), 84–94.

Høgedahl, L. (2019) *Den danske model i den offentlige sektor. Danmark i et Nordisk perspektiv.* Copenhagen: Djøf Forlag.

Jansen, T., Post, S., Kristiansen, T., Óskarsson, G. J., Boje, J., MacKnzie, B. R., Broberg, M., & Siegstad, H. (2016). Ocean warming expands habitat of a rich natural resource and benefits a national economy. *Ecological Applications, 26* (7), 2021–2032.

Marquardt, O. (2005). Opkomsten af en grønlandsk arbejderklasse. In A. Carlsen (Ed.), *Arbejdsmarkedet i Grønland – fortid, nutid og fremtid, 2005* (pp. 176–188). Nuuk: Ilisimatusarfik.

Nuttall, M. (2008). Climate change and the warming politics of autonomy in Greenland. *Indigenous Affairs, 1,* 44–51.

Nuttall, M. (2010). Anticipation, climate change, and movement in Greenland. *Etudes/Inuit/Studies, 34* (1), 21–37.

Nuttall, M. (2017). *Climate, society and subsurface politics in Greenland: Under the Great Ice.* London and New York: Routledge.

Poppel, M. (2005). Barrierer for grønlandske mænd på arbejdsmarkedet [Barriers for Greenlandic men in the labor market]. In A. V. Carlsen (Ed.), *Arbejdsmarkedet i Grønland – fortid, nutid og fremtid [The labour market in Greenland – past, present and future, 2005]* (pp. 125–140). Nuuk: Ilisimatusarfik.

Sejersen, F. (2015). *Rethinking Greenland and the Arctic in the era of climate change: New northern horizons.* London and New York: Routledge.

Skjervedal, A. (2018). *Towards meaningful youth engagement: Breaking the frame of the current public participation practice in Greenland* [PhD dissertation]. Ilisimatusarfik – The University of Greenland and Aalborg University, Denmark.

Stiglitz, J. E. (2010). Freefall: America, free markets, and the sinking of the world economy. New York: W. W. Norton & Company.

Thorleifsen, D. (2005). *Arbejdskraftens sammensætning og spørgsmålet om planlægning og uddannelse i 1950'erne.* In A. Carlsen (Ed.), *Arbejdsmarkedet i Grønland – fortid, nutid og fremtid* (pp. 189–202). Nuuk: Ilisimatusarfik.

2 The Greenlandic economy

Structure and prospects

Torben M. Andersen

Abstract

Scale and distance are key factors for the Greenlandic economy. Greenland has a tiny population with scattered settlements in a huge country with arctic climate and is situated far from potential trading partners. Despite this background, the economic development in Greenland has been favourable, and average material living conditions approach the level in the Nordic countries. However, inequality is high, and the economy depends critically on fisheries and transfers from Denmark, constituting about half of general government revenue. Fisheries is the dominant export sector, and highly specialised, with prawn and Greenland halibut constituting the lion's share of exports. However, efficiency in fishing is low due to more labour and fishing equipment than needed to fish quotas. Regional and employment considerations play an important role for policies. Future developments face several challenges. Welfare arrangements are not financially sustainable given an ageing population. While there is progress, educational attainments are at a low level, and social problems are large. The comparative advantage is natural resources – exhaustible and non-exhaustible – but scale and distance factors constitute binding barriers.

2.1 Introduction

Greenland has a population of about 56,000 inhabitants and therefore clearly belongs to the group of small countries. However, measured by area, Greenland is the 12th largest country in the world (2,166,086 km²), but only about 20% of this is ice-free (410,449 km²). Consequently, the population density is very low; the population mainly lives along the west coast. There are 16 towns and 58 registered settlements. Nuuk, the capital, has 18,300 inhabitants.

Distance matters in Greenland. Situated in the north, distances are far between potential trading partners, but also within the country (the longest distance north to south is 2,670 km; east to west is 1,050 km, and the coastline measures 44,087 km). No roads connect two towns, and transport must take place by sea or air, involving high transport costs in terms of time and/ or money. On top of this, weather conditions can be rough.

In economic terms, Greenland's options are very different from those of most other countries; the keywords are distance and scale. Significant barriers arise from the combination of challenging geography and a small population. Arguably, the economy consists of a number of segmented areas separated by barriers to economic integration and interaction. Despite globalisation, geography still matters, and proximity to trading partners remains an important factor for international trade and thus economic development and growth. The background factors thus include small-scale disadvantages (difficulties in releasing economies of scale) and difficulties in meeting even the most basic principles of a market economy,[1] namely the potential entry of competitors to curtail market power and ensure a competitive and innovative market process.

A direct, simplistic comparison to other countries provides a biased impression of the options and possibilities available in Greenland; the fundamentals are different. Yet, Greenland has to meet the requirements of global markets if it is to develop a self-sustaining economy. The comparative advantages and economic possibilities are tied to exploitation of renewable and non-renewable natural resources. Historically, natural resources (fisheries, hunting, and mining) have been crucial and will remain so in the future, but with increasing focus on the exploitation of non-renewable resources (minerals and fossil fuels) and tourism.

Since the mid-20th century, Greenland has experienced dramatic changes in social conditions and material living standards. Greenland used to be an isolated fishing and hunting society with a low standard of living, but today, it has an average living standard on par with several OECD countries. Housing and health conditions have improved significantly. Over a short span of time, Greenland has undergone a process that, for most other counties, has evolved over a couple of centuries. Such a quick and dramatic change does not proceed without human, cultural, social, and economic problems. Old structures are dismantled and created anew in a process with strong outside influences and input. An assessment of the Greenlandic society and economic structure thus depends critically on whether one focuses on the large, dramatic improvements over a short time period or its contemporary economic and social problems compared to high-income countries.

This chapter contains an overview of the structure of the Greenlandic economy, recent economic performance, and future prospects. The chapter is organised as follows: Section 2 has a brief outline of recent economic developments, and in Section 3, we review the economic structure with particular focus on fisheries and the public sector. Sections 4 and 5 cover population dynamics and education, respectively. We discuss developments in living standards and their distribution in Section 6, and in Section 7, we explore some key challenges in relation to fiscal sustainability and the prospects of making the Greenlandic economy self-sustaining with living standards comparable to those of the Nordic countries. Section 8 concludes the chapter.

2.2 Recent economic developments

Economic developments have been favourable in recent years. Economic activity has been growing and unemployment declining; see Figure 2.1. Development in activity measured by GDP growth underestimates the progress in recent years, as export prices have increased relative to import prices (the terms of trade improvement) (see Section 3.1). Hence, real incomes have grown by more than what the GDP expresses. Increasing activity has been accompanied by increasing employment, and the unemployment level is brought down to its structural level as determined by qualifications, geographical factors, and economic incentives.

From a comparative perspective, it is striking that Greenland did not experience a recession during the financial crisis, as most other countries did. This shows, in a nutshell, that the economy's structure is very different from that of most other countries. Foreign trade is predominantly inter-industrial; that is, imported goods and services differ from exported goods and services. Export is dominated by fish and shellfish (97% of all exports of goods) and concentrated on a few species (of the total export value, prawns account for 48% and Greenland halibut for 28%). Imports are more broadly composed of various raw materials, machinery/equipment, and consumables. Moreover, a more provision-based economy exists in parts of the country. Hence, the usual business cycle mechanisms with strong roles for industrial products and domestic demands do not apply to Greenland. The key foreign impact factor is the price of fish. Public finances are less cyclically dependent due to external transfers constituting about 50% of total general government revenue. Greenland's business cycle dynamics are thus very different from most other countries.

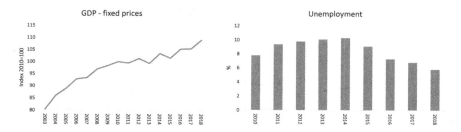

Figure 2.1 Recent economic developments: GDP (fixed prices) and unemployment.

Note: The unemployment metric is defined as persons who within the month have not been in employment.

Source: Statistics Greenland.

Data source: Statistics Greenland, http://www.stat.gl/.

While the COVID-19 crisis arises from a global health shock, the effects are unique in Greenland. On the one hand, the geography makes it easier to implement travel bans and isolation policies, and the spread of the virus has thus been very modest (as of October 2020, no fatalities due to COVID-19 had been reported). On the other hand, reopening travel is entirely dependent on the situation in other countries, since allowing travel may bring the virus into the country. Greenland's health care system is already overburdened and lacks the capacity to handle a serious health situation. Moreover, handling the virus in more distant settlements would be even more difficult. The dilemma is that having the virus under control when the outside world does not makes it difficult to normalise travel, which affects tourism in particular. At the moment of writing, the consequences of COVID-19 have not been dire.

2.3 Economic structure

The economic structure is highly specialised, with fisheries as the dominant sector. Moreover, the public sector in terms of employment is large. Greenland thus shares an economic structure with other economic regions in the Arctic based on natural resources and a large public sector (see e.g., Duhaime & Caron, 2006).

2.3.1 Fishing and hunting

In the fisheries, the three most important species are prawn, Greenland halibut, and cod, for which developments in catch volume and prices are shown in Figure 2.2. Generally, price developments from 2010 to 2018 trended in the upward direction; 2014 catches also increased for these three species. Developments in fisheries have thus been favourable in recent years in both price and quantity dimensions, and this is the main driver underlying recent economic performances; see Figure 2.1. Because of price development, the terms of trade improved by no less than 40% from 2010 to 2019.

The strong dependence on fisheries – and even on a few species – creates a very special situation. The prices of fish are largely determined by the world

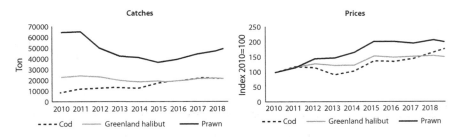

Figure 2.2 Fisheries – prawn, Greenland halibut, and cod, 2010–2019.

Data source: Statistics Greenland, http://www.stat.gl/.

market and thus exogenously given. Stocks in fisheries are determined by biological conditions and management policies. Hence, the economy is quite vulnerable to the world market and biological conditions. While developments in recent years have been favourable, the past has seen many examples of negative shocks originating from fisheries. Notably, from the 1950s to the 1970s, cod fishing was crucial, but the stock dwindled due to overfishing and climatic changes. As a result, the economy ended up in a deep and prolonged crisis.

Fisheries are regulated by quotas. For prawns and offshore fishing of Greenland halibut, quotas are set in accordance with biological recommendations. This aspect of fisheries is certified according to MSC fishery standards, which require sustainability among other factors. Certification makes it possible to charge higher prices. For other aspects of fisheries, quotas are systematically set above biological recommendations; see Greenland Economic Council (Grønlands Økonomiske Råd, 2019). The prime reason for this is regional and employment concerns, as fisheries are the only economic activity beyond the (public) service sector in many parts of the country. The setting of quotas is highly contested. Setting quotas above those specified by biological advice produces short-run gains but at long-run costs in terms of lower incomes and employment. Generally, there is excess capacity in fisheries; that is, more people are employed and more capital is invested in fishing capacity than needed according to the quotas. In short, too many are employed in the fishing sector, which may be interpreted as a form of hidden unemployment. Consequently, productivity in the sector is low, and although there is some scope to expand possible catches of other species of fish, it is not realistic to expect fishing to contribute significantly more than at present; see Fiskerikommissionen (2009). A more effective self-sustaining economy thus requires values to be generated from other forms of economic activity.

However, Greenland is a front-runner in taxing resource rents in fisheries. Such taxation was introduced in 2008 and has since been extended, and it now delivers a substantial revenue corresponding to 10% of total tax revenue. There is an ongoing discussion on fisheries' management, and a commission has been appointed to evaluate and propose changes to the management regime.

Hunting is no longer of major importance for overall economic performance, although it is important in some settlements. Moreover, traditions and cultural values are deeply ingrained in fishing and hunting, and thus of wider societal importance.

2.3.2 The public sector

Nordic countries are well known for large public sectors. Extended provision of welfare services, including education, health, and elderly care as well as a relatively generous social safety net are financed collectively via taxes. In large Nordic countries, the public sector constitutes about 50% of the GDP (cf., Figure 2.3). In Greenland, the public sector amounts to 70% of the GDP.

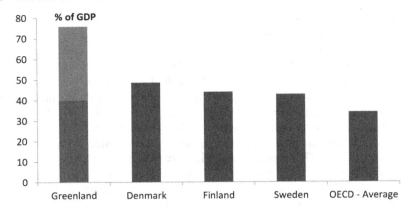

Figure 2.3 Public sector size as a share of GDP; Greenland and Nordic countries.

Note: Public sector size measured by total revenue relative to GDP. For Greenland, revenue is
separated into domestic revenue sources (the bottom part of the column) and revenues
from transfers from abroad (the top part of the column).

Source: Statistics Greenland, http://www.stat.gl/, and OECD, www.oecd-ilibrary.org.

Given location and scale factors, it is to be expected that the public sector
plays a relatively larger role in a country with a tiny, scattered population.[2]
Small-scale disadvantages tend to increase expenditures for provision of
services to the entire population, and it is a particular challenge to be able
to offer all citizens the same opportunities. At the same time, there may
be a need for public-sector activities and involvement due to imperfections
in market mechanisms rooted in small, segmented markets. On top of this
comes the political objective of an extended welfare state of the Nordic type.

In 2018, the total public outlays were about 11.5 billion DKK, of which
public consumption constituted 8.5 billion DKK, social transfers 2 billion
DKK, and public investments 0.9 billion DKK. Notably, social expenditures[3] constituted a smaller relative share than in the Nordic countries.
Obviously, expenditures do not measure the quality and quantity of public solutions. As an example, the cost of providing health care is higher
due to high transport costs and the need to attract foreign personnel; see
Greenland Economic Council (Grønlands Økonomiske Råd, 2019).

The revenue side consists of import taxes of 1.2 billion DKK, income
taxes of 5 billion DKK, capital income taxation and dividends of 1.1 billion
DKK, and transfers from abroad constituting 5.5 billion DKK (including
0.9 billion DKK in reimbursement of expenditures from the Danish state,
3.9 billion DKK in block grants from Denmark,[4] and 0.3 billion DKK from
the EU). Close to half of total public outlays are thus financed via transfers from abroad (cf., also Figure 2.3). Assessed in terms of employment,
about 40% of the employed work in the public sector (Self-Government of
Greenland [Selvstyre] or municipals). The effective number of employees is
likely higher due to outsourcing to Danish companies, for example.

It is worth noting that Greenland does not have a public debt problem. In recent years, fiscal policy has been planned under a budget rule preventing the fiscal budget from relying on deficit financing. The public sector as such does not have a debt (small positive net position), but there is an implicit debt via ownership of companies. In 2019, the gross financial debt was 4.8 billion DKK, and the net debt was 2.8 billion DKK (Naalakkersuisut, 2020). A specific characteristic of Greenland is the large commercial involvement of the public sector. The government owns a number of large companies fully or partially (some of which are monopolies). Examples include (public ownership is given in parenthesis): Royal Greenland A/S (100%), KNI Pilersuisoq A/S (trading and service company; 100%), Tele Greenland A/S (100%), Royal Arctic Line A/S (100%), A/S Boligselskabet INI (housing; 100%), Great Greenland A/S (100%), and Air Greenland A/S (100%). Since most of these companies are engaged in essential activities, they are "too big to fail", implying a significant risk for the public sector (Selvstyret). The governance structure is likewise complicated and challenges the arm's length principle into the political system to ensure that the activities are managed on purely commercial principles.

2.4 Population and migration

The population size has been relatively constant at around 55,000 persons over the last decades. This is the net result of two opposing forces. First, the fertility rate is relatively high, though falling and approaching 2 (the level of population reproduction). Second, in net terms, emigration between locations in Greenland (cf., Figure 2.4) occurs at a level roughly corresponding

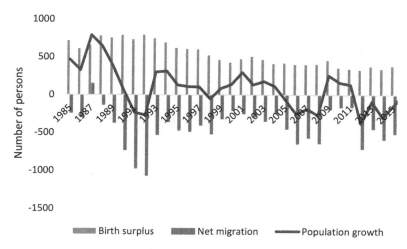

Figure 2.4 Population dynamics, 1985–2019.

Note: Birth surplus is the difference between the number of births and deaths.

Source: Grønlands Statistik, Statistikbanken, Migration (Vandringer), www.stat.gl.

to the birth surplus. The recent projection by Statistics Greenland asserts that the population size is going to decline to about 48,000 persons in 2050, mainly due to declining fertility.

For a small country, it is natural to expect a high level of emigration and return migration for the simple reason that many young people go abroad (mainly to Denmark) to educate themselves. Roughly half the population have spent some period of their lives abroad (Grønlands Økonomiske Råd, 2013). Going abroad for education and to acquire labour market experience is an advantage for individuals and society. However, systematic net immigration can be taken as a sign of underlying problems, as a large share of those migrating do not return to their mother country.

No in-depth empirical studies have analysed migration patterns for individuals born in Greenland. It is often suggested that key reasons for migration are the quality of schooling, health care, and job market prospects. An analysis of those migrating to Denmark shows several remarkable findings (Grønlands Økonomiske Råd, 2013). Return migration rates are much lower for immigrants from Greenland than Iceland and the Faroe Islands. About 80% of immigrants to Denmark from Iceland return to their mother country within 10 years; for immigrants from the Faroe Islands, this rate is 60%, but for Greenlanders, it is only 40%. Moreover, about 80% of those living in Denmark are well integrated into the labour market, with good jobs and incomes. This suggests that net emigration is associated with a brain drain.

The age structure is changing radically. The number of young is declining and the number of old increasing – see Figure 2.5. The flipside is a fiscal

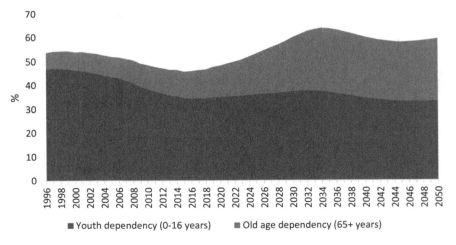

Figure 2.5 Dependency ratio, 2006–2050.

Note: Population aged 0–24 and above 60 relative to age group 25–59. Data for 2013–2050 is
    a projection.

Source: Statistics Greenland, Population projection. www.stat.gl.

sustainability problem; cf. below. The changes in age composition are driven by falling fertility rates and increasing life expectancy, reflecting improvements in living and health standards. For men, life expectancy increased by about 8 years from 1977–1981 to 2014–2019 (from 60.6 to 68.3 years); for women, the increase is about 5 years (from 67.6 to 73 years). Though narrowing, there is still a gap with the Nordic countries. Increases in longevity are expected to continue, and Statistics Greenland predicts longevity to be 72 years for men and 77.3 years for women by 2040. Notably, the mortality rate for men in their 20s is high due to accidents and suicide.

As in all countries, Greenland is experiencing a tendency towards agglomeration alongside economic developments and changes in the economic and social structure of society. While close to 25% of the population was living in settlements in the late 1970s, the share is now down to about 15% (see **Figure 2.6**), and according to Statistics Greenland population forecasts, it will decline further. This process reflects both an adjustment to the geographical location of economic activity and the possibilities offered by bigger towns in terms of education, health facilities, culture, sports, etc. However, in a small population with scattered, separate settlements, the agglomeration process is more challenging than in most counties. For example, in many European countries, smaller villages and towns remain inhabited while people commute to jobs in different areas. This is obviously not feasible in most cases in Greenland.

The agglomeration process is a particularly difficult issue in Greenland due to the cultural importance of settlements and historic experiences with compulsory relocation of inhabitants. With changing economic structures, it is increasingly difficult to maintain job opportunities in settlements,

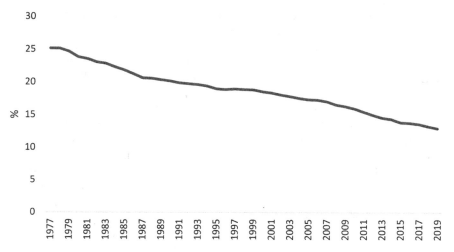

Figure 2.6 Share of population living in a settlement, 1971–2025.

Source: Statistics Greenland. Population forecasts, www.stat.gl.

especially if employers are to offer incomes on par with the bigger towns. At the same time, settlements are rapidly aging as young inhabitants move away, which reinforces the reduction of fertility in these locations. Difficulties in maintaining equal access to schooling may add to this issue, as it is challenging to recruit trained teachers. Likewise, access to welfare arrangements such as health and old-age care is problematic, although some progress has been made via the implementation of tele-medicine. Even access to daily supplies can be a problem, and these are generally more expensive due to transport costs and small markets. On the other hand, settlements carry strong cultural legacies going back to the Inuit hunting and fishing societies. Accordingly, policy questions related to the depopulation of settlements and urbanisation are very sensitive.

2.5 Living standards and distribution

Living standards have improved significantly since the 1950s. Average living standards in 1950 were about half the level in Denmark (see Gad, 1994), while today, the gap has been reduced significantly. GDP per capita in 2018 was about 10% lower than in Denmark.[5]

Taking a wider perspective, living conditions have generally improved. Housing standards have improved significantly over the years, although there are still problems and a shortage of housing (in part due to the agglomeration of the population). There have also been significant improvements in health. However, living conditions are also affected by risk factors such as alcohol, drugs, smoking, and sexually transmitted diseases. Hence, despite the improvements, longevity remains shorter than in most OECD countries (cf. above).

To assess living standards across the population, the distributional profile is important. Table 2.1 reports both the Gini coefficient,[6] as a metric for inequality in the overall distribution of income, and various poverty measures. These measures are compared with the Nordic countries. Both income inequality and poverty are higher in Greenland than in the Nordic countries. There has been a tendency towards larger income inequality and relative poverty in recent years.[7]

It should be noted that the extent of non-market income arising from fishing and hunting is substantial for some groups, implying that registered market incomes underestimate the actual living standards (Naalakkersuisut, 2018). It is difficult to assess the implications for the distribution of income.

The source of inequality is a large dispersion in market incomes. About 40% of employees have an annual wage income less than what corresponds to full-time work at the minimum wage (see Skatte- og Velfærdskommissionen, 2010). There is a close relationship between employment and wages, on the one hand, and education and qualifications, on the other.

Among those falling below the poverty line are many singles, as well as both young and old persons. In general, there is not an overrepresentation

Table 2.1 Income distribution Nordic countries, 2018

| | Income inequality | Relative poverty: Critical income in % median-income | |
		50%	60%
Greenland	35	11.6	17.7
Denmark	28	6.8	12.7
Finland	26	5.4	12
Iceland	25	5.2	10.1
Norway	25	7.1	12.3
Sweden	27	9.3	16.3

Note: Income inequality measured by the Gini coefficient defined over equiv-
alised disposable income. The poverty metric uses the same income
concept. The Gini coefficient is defined to be between 0 and 100%. The
larger the Gini coefficient, the more unequal the income distribution.
Data for 2018 or nearest year.

Source: Nordic Statistics Database, www.nordicstatistics.org.

of families with children. With a poverty line determined by 60% of the median income, 15% of children grow up in poverty; with a 40% limit, the share is 5% (Skatte- og Velfærdskommissionen, 2011).

Most children and youth are thriving and getting a good start on life, but several analyses point to problems for a large group. Christensen et al. (2009) assessed that about 27% of all children are experiencing neglect of care, of which 12% are neglected to a moderate extent and 15% to a serious extent (see also Christensen & Siddhartha, 2015). Moreover, social problems influence parental capacity, and thus in turn hamper their children's development.

Alcohol, drugs, and violence have a predominant role in many homes and families (Skatte- og Velfærdskommissionen, 2011). As an illustration, some 60% of the young have experienced problems associated with alcohol in their family, and 20% of all mothers say that they have experienced physical abuse and threats of violence from their partner. Moreover, there are problems with sexual abuse: About one-third of girls and somewhat fewer boys have experienced such abuse. Suicide or thought of suicide is a significant problem among children and youth. The frequency of suicide for youth below the age of 24 is about 30 times larger than in Denmark.

Finally, the intergenerational linkage in social conditions is extremely strong and reflects itself in a vicious circle with lack of education, marginal attachment to the labour market, early parenthood, abuse, and violence. The social and distributional challenges are thus an important issue.

2.6 Education

Education is a key issue in relation to both the current economic situation and the future development of the Greenlandic economy. Economic development is constrained by lack of qualified labour across most of the educational spectrum. Still, many are unemployed or working on a low income. An increase in qualifications is paramount to improving the economic situation of individuals and families via more stable employment and higher income. This will translate into a more equal distribution of income, as well as a higher level of income in general. Moreover, higher employment rates will have significant effects on the public budget, increasing tax revenues, and reducing social expenditures.

Figure 2.7 depicts the essence of the problem. Figure 2.7a shows employment rates across different educational groups in comparison with Denmark and the OECD average. The data shows the well-known educational gradient that employment rates are increasing with education. The same applies to income, retirement age, and many other factors. Equally important, it shows that conditional on education, employment rates in Greenland are on par with those in Denmark (and thus many OECD countries). Figure 2.7b points to the dilemma arising because a large share of the population does not attain a labour market-relevant education.

Education has high priority, and almost 20% of public expenditures are allocated for this purpose. This is high in a comparative perspective, but there are a number of barriers influencing educational outcomes. The scope for educational activities is hampered by several factors. Location and scale factors make it difficult to offer equal opportunities for all children. In particular, schooling in settlements is associated with problems; there are few children, and it is difficult to recruit qualified teachers. Moreover, a deep-rooted language issue prevails. The official language is Greenlandic, but society is effectively bilingual, with Greenlandic and Danish being used simultaneously. However, not all have a strong proficiency in Danish, and there is also a share of the Greenlandic population that is not proficient in Greenlandic. For historical reasons, there has been a large share of Danish teachers and teaching material, but the share of Greenlandic teachers has increased significantly in recent years. Danish remains important in the educational system, especially beyond primary school, due to access to teaching material, but also

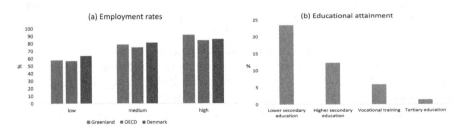

Figure 2.7 Employment and education.

Note: Grouped according to highest achieved education for the age group 30 in 2013.

Source: Statistics Greenland, www.stat.gl and www.oecd-ilbrary.org.

to advance into higher levels of education. A large share of youth goes to Denmark for schooling and studies for a period of time. A recurrent issue is what education should be offered in Greenland and whether students should go abroad. Small-scale disadvantages make it difficult to reach critical mass, regarding both students and teachers, while local education options may be important for intake and the structuring of the programmes. Although the educational system has been expanded, it is not realistic to have a full educational programme offering education at an international level.

A key challenge is primary and lower secondary schooling. Recent analyses have pointed to several problems, also with respect to teacher proficiency. The social gradient is very strong, and the system does not succeed in creating equal opportunity for all. A large fraction of each cohort leaves the schooling system with insufficient proficiency and motivation to pursue further education. In the age group of 16 to 24 years, about one-third is neither employed nor in education.

At all levels of education, there are very high dropout rates, and the number of "retakers" is large. This implies a large turnover in the educational system, which is a further cost driver. There have been no systematic analyses of the reasons for this situation, but weak proficiency in reading, mathematics, and language is among the likely causes. Moreover, education often involves a move from a smaller community to a larger town, which can be difficult. Access to housing is also a constraint. For vocational training, scarcity of "traineeships" and student housing shortage are serious constraints implying that not all qualified applicants are admitted into such educational programs.

The educational system thus faces a number of challenges. Nonetheless, there is progress, and educational attainment is gradually improving (see Figure 2.8). However, from the perspective of supporting a more self-sustaining economy, the speed of progress is too slow. In 2040, the fraction of the labour force without any education beyond lower secondary schooling is, based on current developments, projected to be about 40%.

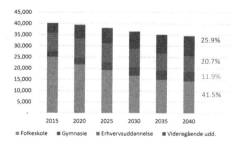

Figure 2.8 Educational composition of the work force – projection.

Source: Grønlands Økonomiske Råd (2019).

2.7 Challenges

Despite the favourable economic developments in recent years, there are several economic challenges. First, the current situation is not sustainable. An ageing population implies a significant financial challenge. Second, ensuring that the economy sustains a higher and more equally distributed level of living standards while at the same time having a more diversified economic structure and a reduced reliance on outside transfers is a particular challenge.

2.7.1 Fiscal sustainability – status quo is not an option

Maintaining status quo is not an option for many reasons. As in many other countries, in Greenland, the ageing population (see above) creates a fiscal sustainability problem. Projections of public expenditures and revenues – for unchanged policies – predict a widening financial gap. Expenditures will steadily increase, while revenue will remain constant as a share of GDP (see Figure 2.9). In short, the current welfare arrangements cannot be financed by the current taxation system (including transfers from abroad) as the population ages. The welfare arrangements are thus not sustainable. Some initial steps, including an increase in the pension age, have been taken, but much more drastic measures are needed to close the financial gap caused by ageing. The government launched a Sustainability and Growth plan in

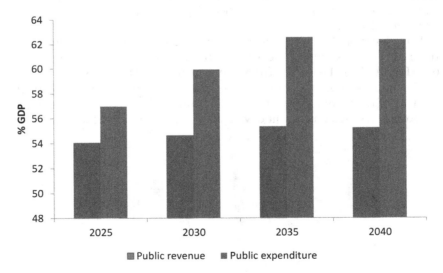

Figure 2.9 Projection of public revenue and expenditure as a share of GDP, 2025–2040.

Note: For assumptions underlying the projections, see Grønlands Økonomiske Råd (2019).

Source: Greenland Economic Council.

2017, which included public savings as a major element to close this fiscal gap. However, so far, no specific actions have been taken to implement the key elements of the plan.

It is worth stressing that the projections reported above do not include any repatriation of activities from the Danish state, and the transfer from the Danish state is included in the projection. Hence, any steps to change any of this require additional initiatives.

2.7.2 A process towards a self-sustaining economy

An essential question is whether Greenland, from a low starting point, is in the process of catching up in terms of improving its production outcome such that living standards are brought up to par with the Nordic countries without transfers from abroad.

One argument in this debate is that the economy is distorted/twisted because of the block grant. Welfare arrangements have preceded the development of an economic base. The economic development has been associated with a large expansion of the public sector (see above) and so-called non-tradeable sectors (housing, services, etc.). In this process, private production and, in particular, export sectors have been squeezed. Shielded by the transfers, there has not been the same urge to develop production possibilities. This might also have muted economic incentives to educate, mobility, etc. Moreover, higher net emigration could be expected.

This resembles the so-called Dutch disease, referring to the possibility that outside transfers or revenue from natural resource extraction crowd out traditional production activities (see, e.g., Torvik, 2009). One mechanism through which this could occur is via wage increase driven by an expanding public sector and non-tradeable sectors. The higher wages hamper competitiveness for export-oriented sectors, which in turn reduces growth. A concern for wage competitiveness has thus not been a major factor in wage determination, but rather the ability to attract workers to activities in the public sector.

Against the relevance of this in a Greenlandic context, it can be argued that there was no thriving private sector to crowd out, and the geography of Greenland (distance and scale) does not make it realistic to have an industrial sector with export potential. Economic conditions are constrained by the available economic resources, notably fishing. It is difficult to make counterfactuals to assess the situation in the absence of economic transfers from Denmark. The most realistic scenario is probably depopulation. The population size increased dramatically in the 1950–1970 period due to the improvements in living standards. Over the years, a recurrent theme has been to promote private enterprises. Already in the 1950s, several attempts were made, but they generally failed, and public intervention was needed. The public sector has thus been extensively involved in production activities, and still is today (cf. above).

The key question is whether a small, scattered population with a challenging geography is able to become self-sustaining. The current discussion focuses on how to broaden economic activities to make the economy both more self-sustainable and diversified, and thus resilient.

Economic sustainability can be defined in many ways. A first and obvious criterion is that public finances should meet sustainability requirements and, similarly, a sustainable use of natural resources. This is in itself a loose definition of sustainability, because this can be attained at different levels of welfare arrangements. Hence, a further part of the definition is that welfare arrangements should be at a satisfactory level. The same applies to living standards; they should be on par with those of, say, the Nordic countries. Moreover, dependence on transfers from abroad should be reduced and eventually eliminated. On the other hand, defining a self-sustaining economy solely in terms of independence of foreign transfers is not meaningful. Declining the transfers is in principle possible, but would require dramatic economic adjustments and cause a significant drop in living standards. This underlines the fact that a self-sustaining economy also includes objectives concerning the living standards achievable for the population.

What is the scope to make the Greenlandic economy self-sustaining? The comparative advantage of the Greenlandic economy is its natural resources – renewable and non-renewable. What are the possibilities here?

Fishery is very important and will remain so. It is, however, not plausible for fisheries to provide a much larger economic base than at present. There is definitely scope for improvements, but that is mainly in terms of efficiency, productivity, and refinement of the value-added chain (see discussion above), which would make resources available for other uses, in particular labour resources, which is a binding constraint for other parts of the economy.

Tourism is often highlighted as having high potential. Nature is fantastic, and there are spectacular options. Currently, tourism is centred on Ilulissat, but there is also some in southern Greenland. After a long debate, a major change in the airport structure is now underway. The airports in Nuuk and Ilulissat are being expanded (in cooperation with the Danish state) to allow transatlantic flights, implying that flight time can be reduced significantly (compared with flying via Kangerlussuaq). An airport is under construction in Qarortoq in southern Greenland (as an alternative to the airport in Narsarsuaq, which is not in proximity to any of the towns in southern Greenland). To expand tourism, investments in hotels and other facilities are needed. The expansion of tourism raises an inherent dilemma between, on the one hand, ensuring scale to allow for reasonable prices and significant economic importance and, on the other hand, the need to preserve the unique natural capital, which is very sensitive, and mass tourism may be detrimental to the attractiveness of these locations.

The prospects of economic development thus depend on non-renewable resources (e.g., minerals and fossil fuels). Historically, mining has been moderately important (e.g., cryolite, coal), and during WWII, the economy was, to a large extent, economically sustained due to mining of cryolite. However, in recent years, mining has not been an important economic factor.

Geologically, Greenland is well researched, and deposits of various minerals have been identified. The potential is, in this sense, large. The challenge is the cost of extracting these resources. Most deposits are located in remote and difficult-to-access areas, and only in few cases is relevant local infrastructure (roads, harbours) available. Mining in Greenland is thus "marginal" in relation to the world market, and therefore the global price developments are crucial for the scope for mining. Recent experience shows that the path from discoveries to commercial exploitation is long and uncertain.

Oil exploration research began in the 1970s, and a new round of exploration peaked in 2010/2011 with some offshore drilling. The outcomes were not encouraging, and these activities are at present on hold. Offshore oil activities are highly debated due to the fragile nature and the particular risk arising due to the climate.

Much focus has been devoted to potential large-scale mining activities, such as the iron mine (Isukasia) and an aluminium factory (exploiting water resources to produce cheap power), due to their potentially large economic impact. However, such activities would be very large in a Greenlandic context, implying, among others, the need to recruit foreign labour and making the economy dependent on a few such large-scale activities and multinational companies. At present, these plans are on hold for various reasons.

Currently, two mines are active: a ruby mine near the Qeqertarsuatsiaat (150 km from Nuuk), employing about 30 persons, and an anorthosite mine at the Kangerlussuaq Fjord with approximately the same number of employees. A number of explorations are in process, and some are approaching implementation. These include rare earth projects near Narsaq in southern Greenland, Tanbreez at Killavaat Alannguat (Kringlerne), and Greenland Minerals and Energy at Kuannersuit (Kvanefjeld). The project at Kuannersuit also involves the mining of radioactive material (including uranium). In 2013, the Government of Greenland abandoned the zero-tolerance policy from 1953 concerning mining of uranium, but the issue remains controversial. Moreover, although the Government of Greenland has sovereignty over natural resources, export of radioactive material also involves issues related to international conventions and foreign policy, which is the responsibility of the Danish government. As of yet, it is unclear whether and when these projects will be realised.

Much focus has been on the government take from possible mining and oil projects. This is clearly important, and there has been extensive discussion on whether this revenue should be collected via taxation or royalties. A resource fund has been established (but is so far inactive), modelled after the Norwegian example, to ensure responsible management of the revenue flowing from such activities. In addition, it must be stressed that the large impact on the economy only arises if the projects result in more activity for local enterprises and create jobs for the population. A large part of the value of such projects is in the extraction activity itself rather than the resource rent. If these projects are not translated into more activity and jobs for the population, it will be very difficult to attain a self-sustaining economy with high living standards and equal income distribution. There are clearly a number of challenges in this respect, including education and housing/commuting. Even if all resource rents are allocated to the resource fund, it is unlikely within a foreseeable horizon to reach a level where the return can compensate for the transfers from abroad.[8]

2.8 Conclusions

The Greenlandic economy is not on a sustainable path, and significant reforms are required. Over-optimistic expectations concerning natural resources and a swift path to a self-sustaining economy have been replaced by a more realistic view on the possibilities (see, e.g., Naalakkersuisut, 2020). Natural resources remain of paramount importance, but the scope depends on both external factors, including world market prices beyond control, and internal factors, which can be affected. The latter includes education, labour market policies, and development of the private sector. Greenland is currently in a position where it must attract foreign labour despite idle local labour, showing that the problem is not shortage of jobs but lack of qualifications, mobility, and job-search incentives. On top of this come looming fiscal sustainability problems due to an ageing population and other factors. There are opportunities and scope for improvements, and some can make an impact on several counts. Improved education will contribute to a decrease in social problems and improve labour market prospects. This, in combination with an expansion of the private sector, will also contribute to better public finances via increased tax payments and lower social expenditures. There is thus room for action to put the economy on a path that eventually could lead to a self-sustaining economy.

Notes

1. It is a unique feature of Greenland that there is no private ownership of land. Land is public property, but individuals can attain land leases, which are basically free.
2. It is not unusual to find that small states have relatively large public sectors, and a significant share of revenue accrues from grants from outside; see e.g., IMF (2014) on Pacific Island Countries.
3. See Chapter 6 in this book and Pedersen et al. (2019) for a discussion of the design of the social safety net and the implication for work incentives.

4. The grant is indexed to price-wage inflation in Denmark. If GDP growth is higher in Greenland than Denmark, it follows that the block grant falls relative to Greenland's GDP. If the expenditure share in GDP is to remain unchanged, other financing sources relative to GDP have to increase.
5. Ideally, the comparison should be based on gross national income (GNI) purchasing power parity (PPP) adjusted, but there are no recent statistical records on GNI.
6. The Gini coefficient measures how far the distribution of income is from a completely equal distribution.
7. In 2002, the Gini coefficient was 34, and relative poverty was 10.6 (50% threshold) and 16.1 (60% threshold; Statistics Greenland, www.stat.gl).
8. See also Udvalget for samfundsgavnlig udnyttelse af Grønlands naturressourcer (2014) for a discussion of the scope mining projects will have for the economy.

References

Christensen, L., Kristensen, G., & Siddhartha, B. (2009). *Børn i Grønland – En kortlægning af 0-14 årige børns og familiers trivsel.* SFI.

Christensen, L., & Siddhartha, B. (2015). *Unge i Grønland – Med fokus på seksualitet og seksuelle overgreb (Rapport 21).* SFI.

Duhaime, G., & Caron, A. (2006). The economy of the circumpolar Arctic. In A. Editor & B. Editor, *The economy of the North* (pp. 17–25). Statistisksentralbyrå Norway.

Fiskerikommissionen. (2009). *Fiskerikommissionens betænkning.* Naalakkersuisut.

Gad, F. (1994), *Grønland, Politikens danmarkshistorie.* Politikens Forlag.

Grønlands Økonomiske Råd. (2013), Grønlands økonomi, Nuuk.

Grønlands Økonomiske Råd. (2019), Grønlands økonomi, Nuuk.

IMF. (2014). *Fiscal monitor April 2014 – Public expenditure reform, making difficult choices.* https://www.imf.org/en/Publications/FM/Issues/2016/12/31/Public-Expenditure-Reform-Making-Difficult-Choices

Naalakkersuisut. (2018). Fattigdom & Ulighed Redegørelse om ulighed og fattigdom samt mulighederne for at fastsætte en fattigdomsgrænse. Departementet for Sociale Anliggender, Familie, Ligestilling og Justitsvæsen.

Naalakkersuisut. (2020). *Politisk Økonomisk Beretning 2020.* https://ina.gl/media/2545676/pkt10_fm2020_pob2020_da.pdf

Pedersen, N. J. M., Petersen, J. S., & Lindeberg, N. H. (2019). *Analyse af offentlig hjælp (Rapport 3).* VIVE. København.

Skatte- og Velfærdskommissionen. (2010). *Hvordan sikres velstand og velfærd i Grønland? Baggrundsrapport.* Grønlands Selvstyre.

Skatte- og Velfærdskommissionen. (2011). *Vores velstand og velfærd – kræver handling nu, Betænkning.* Grønlands Selvstyre.

Torvik, R. (2009). Why do some resource-abundant countries succeed while others do not? *Oxford Review of Economic Policy, 25*(2), 241–256.

Udvalget for samfundsgavnlig udnyttelse af Grønlands naturressourcer. (2014). *Til gavn for Grønland.* Ilisimatursafik og Københavns Universitet.

3 The political economy of Greenland

From colonialism to a mixed economy

Javier L. Arnaut

Abstract

The long-term economic performance of Greenland has been shaped by factors like climate change and global commodity prices but more importantly by the change and persistence of economic and political institutions. This chapter discusses the combination of these factors ranging from natural-resource transitions, historical legacies, and state developmental policies. It provides a general overview to understand the interplay of endogenous and exogenous factors that have influenced the historical evolution of the Greenlandic economy.

3.1 Introduction

The current economic situation of Greenland cannot be properly understood without adequate knowledge of its historical background. Greenland's economic development is deeply rooted in the evolution of the economic relations that originated three centuries ago. Some of the forces that shaped those relations are still present. The existing organisational structure that governs the economy is essentially *path-dependent* from Danish-Greenlandic historical relations. Despite its current legal status as an autonomous country within the Danish Realm, Greenland's economy is heavily subsidised and financially reliant on Denmark. However, this relationship has not been static over time. Instead, it has been a tumultuous and dynamic interplay of exogenous (climate change and natural resource-based commodity prices) and endogenous factors (change of political and economic institutions) that ultimately gave rise to a mixed economy with a large public sector.

The factor endowments (i.e., natural resources, labour, and capital) of Greenland have been structural conditions playing an important role in determining the country's economic orientation. The country possesses unique natural characteristics in comparison with other emerging regions that have faced a transition out of colonialism. Of the Greenlandic territory, 85% is covered by a permanent and uninhabitable ice cap, making the country sparsely populated across coastal communities, with no roads or railroads between them. The absence of market integration and scale economies are central barriers to investment and productivity in many Greenlandic communities. Most of the communities are founded on small formal and informal sectors of traditional subsistence and commercial activities, such as hunting and fishing. Notwithstanding the barriers, Greenlandic labour possesses important qualities characterised by its adaptability to social and environmental changes.

Many Greenlanders consider their traditional economic activities such as sealing and whaling to be a vital part of their cultural heritage. Institutions and policies during the 19th century deliberately preserved the prevailing Greenlandic historical heritage, looking to maintain the trade monopoly, and imposed social regulations channelling the resultant local economic profits to sustain the colonial administration (Rud, 2017). Some informal institutions from the colonial era persisted after colonialism, which continued to influence the economic orientation of the country (Petersen, 1995). An era of state-developmentalism after World War II characterised by large capital investment imposed social and economic interventions that were considered necessary to encourage economic growth and improve living standards. Thereafter, the type of economic institutions persisted, but political institutions were transformed when Greenland transitioned into self-governance. Thus, the evolution and current situation in Greenland cannot be solely linked to the peculiarities of the country's geography and factor endowments. Greenland's long-run productivity has been influenced by persistence and change of institutions over time.

At first glance, the long-term development of Greenland appears to follow Acemoglu et al.'s (2002) "Reversal of Fortune" hypothesis of comparative development that explains income differences of former colonies as a result of differences of institutions. To some observers, the idea that previously poor and sparsely populated territories featured colonial settlers that transplanted "good" institutions leading to higher incomes today (and vice versa for densely populated and previously rich areas) could be a suitable generalisation of Greenlandic historical development. After all, Greenland's current level of national income is relatively higher than that of many other former colonies (e.g., Latin America and Africa). However, that perspective not only ignores that the Greenlandic economy is not yet financially self-sufficient (and that gross income is biased by an annual fixed transfer from Denmark) but more notably disregards specific institutional changes and persistence across time that have transformed the country into a mixed economy carrying structural economic issues.

It is then important to emphasise that the long-term performance of Greenland originated from a combination of endogenous and exogenous factors ranging from historical legacies to developmental policies, climate change, and the labour tensions between distant administrators and Greenlanders in their quest for legitimacy. The country has experienced natural-resource transitions, from being a whale and seal hunting-based economy to a codfish powerhouse and subsequently to an economy based on shrimp fisheries. A new transition into other sectors (i.e., mining and tourism) is underway; however, it carries significant barriers that stem from a similar interplay of historical institutional factors. Although a heavy reliance on the public sector has been an effective mechanism to stabilise the economy and generate income, excessive dependence could be detrimental if public revenues become closely tied to the volatility of commodity markets, as well as if this over-reliance on the public sector hinders the emergence of the local private sector. This chapter broadly discusses these transitions, emphasising the interactions between economic and political institutions, historical legacies, and the general transformation of the Greenlandic economy from a long-term perspective.

The chapter is structured as follows: The next section addresses the colonial origins of the Greenlandic mixed economy, followed by a brief discussion on the transformation generated by climate change and the interwar period. Issues from the modernisation era and the developmental state are discussed in a subsequent section. A brief analysis is devoted to aspects of autonomy, institutional change, and persistence, followed by conclusions.

3.2 The colonial origins of the Greenlandic mixed economy

Greenland is a mixed economy because it relies mainly on public consumption emanating from a large public sector that interacts with state-owned companies operating within *quasi*-markets and a small private sector. It is also a mixed or a *dual* economy from the production side given that there is a combination of subsistence and commercial wage activities that have coexisted for centuries (Caulfield, 1993; Poppel, 2006). However, the existing mixed configuration and coexistence of wage and non-wage activities are the result of a variety of endogenous factors, such as colonial institutions, and exogenous factors, such as climate change and global commodity prices.

There is still considerable debate among academics about the role of colonialism in Greenland. Apologists of the Danish colonial rule argue in general that Greenland benefited from the training and educational systems and physical capital such as harbours and buildings erected by the missionaries and colonial administrators. On the other hand, some academics argue that Greenlanders were exploited by a colonial regime and deprived of determining their own future because the decisions were made by missionaries and traders in the self-interest of Denmark (e.g., Petersen, 1995; Petterson, 2012). Regardless of the perspective adopted, there is an undisputable dimension of extractiveness of the policies in place during the colonial rule, which had a profound impact on the way the economy was organised.

As with other colonisation experiences, the incentive to colonise Greenland was to extract economic surplus. Although early on, the desire of the missionaries to evangelise the indigenous Inuit was fundamental, the financial gains from trading products in Arctic waters were thought to be considerably large, and thus they became a priority to sustain the mission and to consign the surpluses to the Kingdom of Denmark (at the time the Dano-Norwegian Union). Trade accounts from the late 19th and early 20th century show that significant trade balances with Greenland were recorded in favour of the Danish state (Sørensen, 2007, p. 15). The birth of the Greenlandic colonial trade has been widely credited to the arrival of the Dano-Norwegian missionary Hans Egede into Nuuk in 1721. Financed by the Crown, Egede had formed The Bergen Company, which had been granted a monopoly with Greenland. However, his efforts and other subsequent attempts (e.g., General Trade Company from 1747 to 1774) at establishing a profitable mission were unsuccessful.

An effort to permanently establish and maximise the profitability of colonial trade emerged with the creation of The Royal Greenland Trading Department, or KGH (for *Den Kongelige Grønlandske Handel* in Danish), in 1776. The production and trade were focused on the by-products of sea mammals, such as furs and skins, meat, blubber, whale, and seal oil from West Greenland. The company would purchase the products locally and export them (with a significant price margin) to Denmark and import consumer goods to sell them (under a credit-barter system) within Greenland. In 1787, the King's *Instrux* (Instructions) gave authority to the KGH to manage the country's public administration. The company became essentially a *monopsony* until the early 20th century.

By 1800, the European demand and prices for lamp oil and furs were on the rise, and thus, other major trading countries (i.e., the Netherlands, United Kingdom, and Germany) considered the by-products of whale and seal to be highly valuable commodities playing an import role in European industries and households. The KGH sought the market opportunity to significantly expand production by increasing sealing and whaling and maximising the profitability of trading these commodities across Europe. Kayak hunters in Greenland opposed to large-scale production by refusing to use newly introduced guns for seal hunting. They claimed that although the catch was higher, the loss rate was too high, as a proportion of the seals killed sank. The colonial administration overrode the claims of labourers, whose bargaining power was meagre or non-existent. Several decades later, the introduction of new techniques using motorboats helped to solidify the technological restriction (Kapel & Petersen, 1986).

The local prices of Greenlandic products were not set by market mechanisms. KGH fixed equal prices for products, as well as for imported products, irrespective of their internal costs of transportation. While margins of exports usually hovered around 1:7 of their value in Europe (Marquardt, 1999a), in periods of low catch and scarcity of local foodstuffs, the company had to pay workers with imported foodstuffs. Although this situation generated economic exposure for workers given that their labour earnings did not vary according to their productivity, the company was also vulnerable in periods of low economic activity.

However, corporate control over Greenlanders was not exclusively directed to regulate economic relations. The regulation encompassed aspects of social relations, gender, and reproductive rights (Seiding, 2013). The regulation of public life aimed at imposing the Western social values of the time, looking to align labour with the economic interest of the company. The figure of the male master hunter (*Piniartorsiaq*) was highly praised by the KGH and thus started to be seen as more important socially than the previous collective-shared work and egalitarian relations between men and women prevalent in precolonial Greenland. In the 1820s, Greenlandic men began to receive privileged training as administrators, church catechists, and KGH staff employees. Some women were trained as maids (*kiffat*) to work exclusively for Danish families, and they often became wives of the educated elite (Arnfred & Pedersen, 2015).

The second half of the 19th century was an era of important institutional change. Pressured by the shortage of goods for trade, in 1835, the Danish Crown instructed the KHG to pay Greenlanders in cash. Moreover, influenced by a wave of liberalism in Denmark, the policies of the KHG became more flexible, allowing Greenlanders to trade consumer goods locally (Marquardt, 1999a). According to correspondence between two inspectors at the time; "it must be left to the discretion of the native-born to ... set the proper limits to their consumption of European goods" (Marquardt & Caulfield, 1996, p. 112).

The new cash economy and flexibility of domestic trade had an enormous impact on the general consumption of the local population (Petersen, 1990). Previously, Greenlanders would prepare large food provisions for the winter (drying meat and fish), but because cash facilitated immediate exchange at any time, there was less of a necessity for precautionary in-kind savings. However, the tradition of sharing goods without a monetary exchange remained central among some Greenlandic households (Petersen, 1989). By the 1850s, the new flexibility was not well perceived by Danish traditionalists and other colonial officials living in the country who believed that the flexibility would disintegrate society as the Inuit began to exchange traditional goods for imported "luxury" goods such as coffee, sugar, and white bread. The general concern was that the new generation of labourers would get "deskilled" out of traditional activities (e.g., building kayaks and seal hunting) and lured by the trade of luxury goods or fishing (Marquardt, 1999b).

Various Danish colonial officers (e.g., Hinrich Rink, inspector of South Greenland) and traditionalists proposed the restoration of Greenlandic traditions and a return to the original pathway of the domestic trade. A series of reforms were introduced to the system in 1857. The most relevant reform was the introduction of local councils formed by a board of guardians (*Paarsisut*). Only successful seal hunters were elected to the boards, and they acted as the "local eyes" of the colonial authority watching over the conduct of the Greenlanders (Rud, 2017, p. 38). These local boards of guardians were introduced across the country between 1863 and 1867 and played a key role

Table 3.1 The occupational structure of Greenlanders, 1834–1901 (in %)

	1834	*1860*	*1890*	*1901*
Hunter	86.2	82.4	87.9	72.4
Fishermen	-	-	-	14.6
Mission, Church, School	8.2	9.1	3.8	4.7
Administration	0.0	1.1	1.8	0.0
Trade	5.6	7.4	6.3	6.2
Other	-	-	0.2	2.1
Total (%)	100	100	100	100
Total persons employed	**1,290**	**1,971**	**2,407**	**2,749**

Source: *Folketællingen i Grønland 1901*. Data for the years 1834 and 1860 are from *Statistisk Tabelværk* and *Statistiske Meddelelser*, respectively.

in the allocation of local resources. Tariffs were imposed on the domestic sale of coffee, sugar, and white bread, and monetary bonuses (known as repartition) were distributed among the best hunters (Tejsen, 1977, p. 460). Years later, the same boards proposed including fishermen in the system of repartition; however, the KGH rejected this proposal (Andersen, 1999).

Table 3.1 depicts how hunting activities dominated the occupational structure of Greenlanders over other occupations prior to the 20th century. Occupational shares in trade declined, and although fishing was a predominant subsistence activity across settlements, official censuses have no record of employment in fisheries before the start of the century, when fish prices began to be listed in the KGH. In 1864, the Danish state allowed the operation of a private company to mine cryolite in the town of Ivittuut (on the south-west coast of Greenland). However, all the mining workers and managers were brought from Denmark, and no contact was allowed between them and the local population (an isolation policy of the company that continued until the mid-20th century). The mining company paid the state royalties, which were at the disposal of the KGH to finance the company's budget during periods of deficit (Sørensen, 2007).

Historically, fisheries had been an important part of traditional subsistence, but for Danish traditionalists (and originally for the KGH), fish was not a valuable commodity. Fishing was perceived as the trade of the "weaklings", reserved for women and for disabled men who could not hunt (Langgård, 1999). Thus, the reforms of the 1850s can be considered repression of an imminent economic change. These reforms were disguised as an inclusive restoration (because Greenlanders were part of the boards) of traditional activities. In general, the second half of the 19th century is considered a period of socio-economic crisis in Greenland (Marquardt, 1996, p. 88). The reforms of the 1850s that discouraged local trade and the development of fisheries proved to be financially unsuccessful in the decades to come. The international price of seal and whale oil began to rapidly decline as the use of competing energy commodities such as coal and oil spread widely across industrial Europe. The isolationist colonial policy that encouraged a hunting-based economy was simply incompatible with the transformation of the global economy.

3.3 An unintended transformation: climate change and the interwar

The first half of the 20th century was a period of exogenous shocks: climate change and the World Wars. This series of events dramatically changed the orientation of Greenland's economy. In 1908, the KGH was relieved of its administration duties in Greenland's civil affairs. The formal introduction of the administrative entity of municipalities (*kommuner*) and the opening of an office in Copenhagen to manage trade were key developments. However, these changes in domestic policy were tied to a major exogenous "catalyser" of economic transformation: climate change. A rapid rise of temperatures in the North Atlantic Ocean caused a decline of seals (migrating farther north) together with a massive influx of codfish along the Greenlandic coasts (Rasmussen & Hamilton, 2001). The shock meant that the long-standing policy of encouraging the preservation of traditional hunting had to be relaxed, and schemes for the development of commercial fishing had to be formulated.

Although small-scale commercial fisheries existed in Greenland before this phenomenon (factories were built in Ilulissat in 1904), medium- and large-scale fishing was established around the first two decades of the century across coastal communities looking to take advantage of the natural resource windfall caused by climate change. In an early stage of this development, foreign halibut fisheries were built near the town of Sisimiut (centre-west) and Qaqortoq (south). In a later stage, various processing plants for salted cod were built along the west coast (Rasmussen, 2000). Figure 3.1 shows the global importance of Greenland in the fishing industry during these years. The country became a major international player in the codfish industry. Although the years after the Great Depression affected Greenland's fishing trade, by the year 1947, Greenland controlled 45% of the global catch.

The fishing industry profited largely from global demand, even during the interwar years. World War I generated supply shortages in Europe and thus the international price of fish and other maritime species rose during the interwar period (except during the Great Depression). However, although many Greenlandic fishermen benefited from the fishing boom, in general, Greenlandic civil society was still a bystander in this development because the country's economic policies were managed from Copenhagen. The situation changed with the outbreak of World War II. Greenland became relatively independent due to the German occupation of Denmark, meaning that the Danish monopoly in Greenland could not be maintained. As a result, Greenland experienced the benefits of high prices from the "free market" and doubled its cod landings (Holm, 2012).

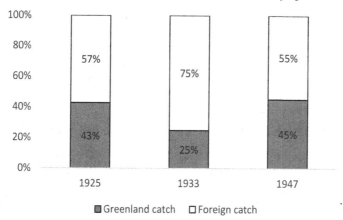

Figure 3.1 The global fishing distribution of codfish catch, 1925–1947 (percentages).

Note: Relatives shares to the total world catch of Atlantic cod (*Godus morhua*).

Source: Derived from data by Rasmussen and Hamilton (2001).

World War II also changed the local Greenlandic perception of their exist-ent economic conditions and growth potential. Within Greenland, in those years, the United States was considered the only country with the ability to provide essential supplies. In 1941, Denmark and the United States signed a defence agreement that allowed American involvement in supplying goods and protecting the island. Because Greenland had the only mine in the world producing cryolite, which was used to produce aluminium, a vital metal for the fabrication of military aircraft, this natural resource advan-tage (and the strategic geographical location) was used as leverage with the United States. Looking to maintain the recognition of Danish sover-eignty over the island, the negotiation was conducted independently by the renowned Danish diplomat Henrik Kaufmann (Heinrich, 2017). However, Greenland's contact with the United States through media and imported products (American and Canadian consumer goods) had a critical effect on the local population. Many began to recognise the difference between their own concept of modernity and modernity abroad. This understanding made Greenlanders realise that their material conditions could be improved if the country were more open to the outside world (Beukel, 2010).

After the war, the Danish authorities looked to normalise relations with Greenland. However, local Greenlandic leaders challenged the imminent return to the pre-war status quo and proposed the creation of a single gov-erning council in Greenland that could systematically plan the local eco-nomic development every five years. The underlying intent of changing governance according to local leaders such as Jørgen C. F. Olsen, a pop-ular Greenlandic politician from Sisimiut, envisioned further involvement in the decision-making to build up Greenlandic private capital and design in a more balanced way through various investment projects in settlements and towns (Fleischer, 2000). On the other hand, Danish administrators were sceptical about whether Greenlanders could manage a liberalised and decentralised economy (Nuttall, 2005). Ultimately, the Greenlandic proposal was rejected in Demark, and in 1948, Danish politicians opted to appoint a "Big Commission" formed by Danes and Greenlanders (Caulfield, 1997).

3.4   Modernisation and the developmental state in Greenland

The Greenlandic initiative from the interwar period to gain greater administrative control did not vanish. On the contrary, the international political environment pressured domestic politics (in Denmark), which tilted the scale towards the notion of self-determination. At the end of the war and under the IX chapter of the United Nations (UN) charter, Denmark was required to submit annual reports to the UN on Greenland's decolonisation efforts. This symbolised a step towards accountability and greater involvement of Greenlanders in the country's development. Moreover, the legal status as a colony ended in 1953 with the adoption of the new Danish Constitution. In principle, the new status of Greenlanders meant an era of equal citizens within the Danish Realm. However, according to Dahl (1986), in reality, for some Greenlanders, it meant an era of labour displacement as the number of Danes relocating to Greenland increased from 4.5% to nearly 20% of the country's population in only two decades (1950–1970). The inbound Danish workforce frequently took well-paid jobs in the public sector and construction.

The post-war period also signified an era of large-scale modernisation based on capital-intensive investment. A series of new reforms, such as the abolition of the KHG monopoly, looked to design a comprehensive plan for economic development. The commission that had been set up in 1948 oversaw a set of policy recommendations to implement a 10-year plan for the country's economic development. The plan was published in 1950 and since then has been commonly known as the G-50. The plan from the commission had two key guiding principles: first, that private capital should lead to a significant increase in production, and second, that income transfers from Denmark should ensure a further increased standard of living, primarily by using the money transferred for the expenses of public consumption (Poppel, 2005, p. 697).

The G-50 was an attempt to launch Greenland onto a self-sustaining growth path, similar to the assumptions of the "Big Push" model of the developmental state (large and simultaneous sectoral investments and trade protectionism). This strand of economic thought was widely accepted and considered appropriate for the industrialisation of developing regions around that period (e.g., Easterly, 2006). The Big Push in Greenland injected large amounts of capital in select sectors such as housing, health, education, and fisheries. The investment in physical capital was mainly channelled to fishing production plants, fishing vessels, power plants, hospitals, and training facilities. An important element of the plan was the attraction of private capital. However, at the end of the 1950s, private investment represented only a marginal share of total investment (public and private), creating concerns about long-term financial sustainability.

The planners' dissatisfaction with the G-50 due to the low levels of private investment led to the creation of a new commission that designed the ensuing G-60. A crucial element of the G-60 was the relocation of the Greenlandic workforce to relatively larger towns where the fishing trawlers were located. Relocating (or resettling) labour into larger towns looked to increase efficiency from *agglomeration* economies, where the immediate availability of local labour would lower search costs, generating higher productivity and further economic expansion and urbanisation.

The policy intervention took place by encouraging inhabitants of small settlements to move to medium and larger towns, particularly Nuuk, the country's capital. The resettlement was not forceful, but Greenlanders were coerced to move because the settlement's public utilities and state-owned stores were closed by the authorities. As a result, settlements began to depopulate, and many workers from traditional subsistence activities were reluctant to move, which generated social conflict. Paradoxically, one of the earlier demands by local Greenlandic leaders in the aftermath of the war emphasised a more balanced investment between settlements and towns. The opposite occurred from 1960 onwards. Figure 3.2 displays the vast and increasing disparity over time of public investments between settlements and towns.

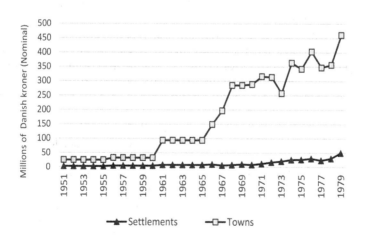

Figure 3.2 Public investment in Greenlandic settlements and towns, 1951–1979 (millions DKK).

Source: Adapted from estimates by Rasmussen and Hamilton (2001).

Note: Figures are expressed in nominal terms.

Despite the seemingly undesired trend of public investments, to some observers, the plans of the G-50 and G-60 accomplished fundamental goals. During this period, Greenlanders' income increased substantially (2.5 times in real terms), and investment in the health sector reduced mortality rates, particularly infant mortality (Kleivan, 1984, p. 703). Greenland was transformed into a high-growth export-oriented economy with strong capital deepening. However, social and economic disparities among Danes and Greenlanders began to appear and became increasingly noticeable by the end of the 1960s. Estimates by Kleivan (1969, p. 146) indicate that 50% of the country's private income in 1967 was earned by Danes living in Greenland.

Along with high unemployment rates, large salary gaps emerged between Danes and Greenlanders. Some of these disparities were accentuated as a result of a national salary rule that promoted Danish immigration into the country. The commission voted for the implementation of a scheme where salaries in the country were set according to employees' place of birth. The so-called birthplace criterion (*fødselskriteriet*) established that foreign (mainly Danish) employees were legally entitled to higher salaries, bonuses, and better working conditions than were employees born in Greenland in the same occupation (Poppel and Stenbæk, 2005).

The country had gone through a hard process of state-developmentalism with interventions that planners perceived as necessary to promote self-sustained growth. The interventions, however, undermined social and governance implications. The economy transitioned into a growing fishing powerhouse carrying major structural issues that materialised in income inequality and a strong dependence on imported human capital and basic goods. The apparent exhaustion of the model did not translate into a regime change. Income transfers from Denmark and high-output growth rates had formed a welfare state that delivered higher living standards. However, the natives felt they were in the backseat of this development. Although there was a desire for greater economic freedom, the key pressure for change was on local governance and self-determination. It was believed that the economic model could be "fine-tuned" under the governance of Home Rule.

The massive local discontent with the deliberate labour market discrimination and the lack of social integration from the settlement-to-town relocation sparked tensions and political mobilisation. Major political parties were formed during the 1970s (Party Siumut and Inuit Ataqatigiit). The anti-colonial sentiment intensified after a 1972 referendum, when Denmark voted to join the European Economic Community (EEC), and although Greenlanders voted against it, the country was obliged to join given the absence of Home Rule. Local politicians believed that joining the EEC would imply less Greenlandic control over the country's fisheries management and thus, additional external regulation from Brussels was considered another unfair imposition. In 1973, a committee was created to explore the possibilities of Home Rule preserving the unity of the Danish

Realm. 1 May 1979 represented a milestone for political change in the country when Greenlanders voted in a consultative referendum in favour of the Home Rule.

3.5 Autonomy, institutional change, and persistence

The Home Rule transferred the legislative and administrative powers to Greenland. Although under Home Rule Denmark retained authority over important areas, the responsibility of the economy (except for monetary policy) and infrastructure projects fell into Greenlandic hands. However, the Danish modernisation project and the model of a developmental state continued. The continuation was attributed to the Greenlandic politicians' hope that by maintaining the prevailing model, the living standards of the country would "catch up" to a level comparable to that of Denmark (Larsen, 1992). An exogenous factor appeared in the early 1980s impacting the economy similarly to the supply shock at the start of the century when the total seal catch declined and the codfish catch increased. This time, the change in sea temperature caused a drastic decline in the stock of the Atlantic codfish population. Overfishing also exacerbated the rapid fall in the codfish stock. The decline occurred in parallel with a marked increase in the shrimp stock (Hamilton et al., 2003).

Aside from the losses from the external shock on the volume of the export sector and the inherent costs of technical change of industry restructuring from codfish to shrimp during the 1980s, the new era of autonomy faced important endogenous issues remnant from the modernisation policies of previous decades. The marginal role of private capital in the share of total investment together with market risks represented (until today) a key hurdle for sustained economic growth. Apart from the small market size, one of the factors preventing the expansion of private capital is the organisational dependence and institutional inertia (Jonsson, 1996). Because of the over-specialisation of fishing activities and the long-standing reliance on the process management systems from Denmark and its skilled workforce, existent business networks in the country consolidated their organisational links solely with Danish firms. As public contractors in Greenland maintained their networks with Danish suppliers, the potential cost-reduction gains and knowledge spillovers from closer competitive suppliers from Canada, the United States, and Iceland did not materialise.

In 2009, the Act of Greenland Self-Government was adopted, and the country gained further control of its natural resources and other key areas. Although the change of political institutions was another historical watershed for the self-determination of the country, important economic issues remained. One of the remaining vulnerabilities is excessive economic reliance on the public sector. The large size of the public sector has been an effective mechanism in generating income. However, although this dependence is not exclusively detrimental (job security from public sector employment acts as insurance against undiversifiable external risk), excessive reliance on the public sector generates other economic issues. The public sector may not crowd out the small private sector of Greenland, but it may affect its sectoral composition. Expansion of a large public sector in a small economy generates an increase in the demand for local private services and products, which might "trap" the overall structure in similar economic activities (non-tradeable sectors). Additionally, issues of fiscal sustainability can arise during the economy cycle if there is little or no sectoral diversification to expand the fiscal base.

Table 3.2 displays the occupational structure before the Home Rule and in 2018 (after the Self-Rule). The large size of public administration has been

Table 3.2 Occupational structure in Greenland in 1976 and 2018 (percentage of total persons employed)

Sector	1976		2018	
	Persons born in Greenland	*Persons born outside Greenland*	*Persons born in Greenland*	*Persons born outside Greenland*
Agriculture, fishing, hunting, and related industries	16.7	1.8	19.0	9.2
Mining and quarrying	0.2	0.1	0.4	1.9
Manufacturing	16.5	8.6	3.4	1.3
Public utilities	1.2	3.0	1.4	2.6
Construction	9.4	22.9	6.7	12.7
Wholesale and retail trade	12.6	7.7	13.4	9.3
Transportation and storage	8.4	10.3	7.1	7.5
Services	7.8	7.4	8.8	17.9
Professional, scientific, and technical activities	7.5	15.5	1.3	6.2
Public administration and service	19.7	22.5	38.6	31.4
Total	**100.0**	**100.0**	**100.0**	**100.0**

Source: Elaboration based on data from the annual report by Ministeriet for Grønland (1981) and Statistics Greenland (StatBank accessed in 2020).

Note: Branches were re-classified following the ISIC 2.0 and the previous aggregation from the employment census of 1981. The branch of services refers to financial and insurance, accommodation and food, information and communication, and real estate activities. Manufacturing includes tanning and dressing of leather, dyeing of fur and manufacture of wood, and manufacture of other non-metallic mineral products.

a common denominator over time; however, the latest share of employment in the sector is a significant increase (from persons born both inside and outside Greenland) relative to the era before the Home Rule. The table also indicates how the change of political institutions (Home Rule and Self-Rule) through political power ("de facto") was used to alter the existing employment allocations by birthplace (i.e., persons born in Greenland took over larger sectoral shares than did persons born outside the country) without altering the general employment distribution. Although there have been some notable sectoral changes (for persons born outside Greenland), the overall economic structure characterised by the prevalence of the public administration and the primary sector has remained unaffected.

Table 3.2 might suggest that after more than four decades, the economic institutions (i.e., organisation of markets, property rights, etc.) from the Danish modernisation era remained relatively unaffected by the changes in political institutions (i.e., rules affecting legislative bargaining, constraints on the executive, etc.). Although the Home and Self-Rule governance brought a historic change in the ethnic and cultural identity of the country and its political class ("de jure"), the persistence of these economic institutions remnant from previous decades is possibly related to the persistence of the distribution of political power under the new governance. Like other international experiences of democratic transition, after political change, the new distribution of political power lies within certain elite groups that prefer to maintain the same type of economic institutions because they can influence the distribution of economic gains (Acemoglu & Robinson, 2008).

However, the new government under the Self-Rule desires economic independence because it is envisioned as the last step of decolonisation as it has a legal base in the Self-Government Act of 2009. This aspiration for further autonomy is transforming the political economy of Greenland. Local policymakers have recognised the urgent necessity to promote economic growth via sectoral diversification. However, the economy is bounded by labour and capital shortages, which are significant bottlenecks to diversification. High economic growth rates are essential to attain financial self-sufficiency to cover for the Danish block grant (a fixed annual transfer from Denmark's public finance); thus, policymakers have targeted mineral mining exportation, agriculture, adventure tourism, and transportation infrastructure as important industries for further promotion.

The new industrial strategy that intends to deliver high economic growth has begun to be questioned by environmentalists because it is believed to significantly raise CO_2 emissions, which could be detrimental to the country's fragile natural environment (Bjørst, 2018). Additionally, although warmer temperatures in the Arctic are likely to uncover areas rich in mineral deposits, offering more opportunities for resource exploitation (e.g., Dale et al., 2018), there is no certainty of the positive spillover effect on the economy. Still, many Greenlanders and politicians see the warming Arctic as an economic opportunity. Although the country complies with the UN

Framework Convention on Climate Change, in 2016, the Government of Greenland requested Denmark to make a territorial reservation against the Paris Agreement. The main argument against the climate agreement is associated with the lack of a binding reference to the rights of indigenous peoples or indigenous peoples' right to development. Efforts to promote sustainable development are also part of the agenda. Currently, the country is among the world leaders in renewable energy given that more than two-thirds of total electricity generation emanates from hydropower.

Compounded by the effects of climate change, it appears that there is another natural-resource transition underway for Greenland. Due to a rise in global demand and supply shocks, and after a long slowdown, the international value of various mineral commodities with which the country is endowed (i.e., iron, lead, zinc, gold, gemstones, rare-earth elements, and uranium) has begun to increase (Humphreys, 2019). Mineral commodities markets are inherently unstable. Reliance on tax revenues from mineral mining in the long term might not be a full remedy against uncertainty. When public budgets are not insulated from market volatility, major budget cuts occur, and social investments cannot be properly implemented under "stop-and-go" spending. The country's political and economic institutions will be tested on whether they are sufficiently advanced to look for new sustainable sources of economic diversification to endure the vulnerability of the new transition.

3.6 Conclusions

From a long-term view, Greenland has experienced a gradual process of institutional change. First, the economic institutions that transferred the local economic gains into the Danish colonial administration were gradually transformed at the beginning of the 20th century, while the country's political institutions remained rather intact. The factors that triggered these economic institutional changes stemmed from a natural resource transition created by climate change and the rise of the international value of fisheries. Second, the interwar fishing boom and the abolition of the trade monopoly at the end of the Second World War produced the seeds of a political institution transformation that eventually materialised at the end of the 1970s. Third, the developmental state policies from the post-war period to the early 1970s created a welfare state and a large public sector that generated and allocated (unevenly) significant economic rents. Fourth, after 1979, an additional natural-resource transition occurred within the fishing industry (from codfish to shrimp) and a major change of political institutions came through the new governance of the Home Rule and Self Rule, yet the economic institutions from the era of the developmental state persisted.

The desire for economic independence is unfolding an imminent transformation in Greenland's political economy. The transformation is an endogenous factor encouraged by exogenous factors such as the natural-resource transition, which is heightened by climate change and the rise in global demand and supply shocks of the mineral commodities. The success of harnessing the new transition's benefits may rely on the ability to transform the country's economic institutions towards the type that promote a more diversified and environmentally sustainable structure.

References

Acemoglu, D., Johnson, S., & Robinson, J. A. (2002). Reversal of fortune: Geography and institutions in the making of the modern world income distribution. *The Quarterly Journal of Economics, 117*(4), 1231–1294.

Acemoglu, D., & Robinson, J. A. (2008). Persistence of power, elites, and institutions. *American Economic Review, 98*(1), 267–293.

Andersen, V. (1999). The impact of commercial fishing on household structures in Sydprøven. In O. Marquardt, P. Holm, & D. Starkey (Eds.), *From sealing to fishing: Social and economic change in Greenland, 1850–1940 (Studia Atlantica)* (4th ed.) Fiskeri og Søfartsmuseet.

Arnfred, S., & Pedersen, K. B. (2015). From female shamans to Danish housewives: Colonial constructions of gender in Greenland, 1721 to ca. 1970. *NORA-Nordic Journal of Feminist and Gender Research, 23*(4), 282–302.

Beukel, F. (2010). Greenland and Denmark before 1945. In F. Beukel, P. Jensen, & J. Rytter (Eds.), *Phasing out the colonial status of Greenland, 1945–54: A historical study*. Museum Tusculanum Press.

Bjørst, L. R. (2018). The right to sustainable development and Greenland's lack of a climate policy. In U. Pram Gad & J. Strandsbjerg (Eds.), *Politics of sustainability in the Arctic* (pp. 120–135). Routledge.

Caulfield, R. A. (1993). Aboriginal subsistence whaling in Greenland: The case of Qeqertarsuaq municipality in West Greenland. *Arctic, 46*(2), 144–155.

Caulfield, R. A. (1997). *Greenlanders, whales, and whaling: Sustainability and self-determination in the Arctic*. Dartmouth College Press.

Dahl, J. (1986). Greenland: Political structure of self-government. *Arctic Anthropology, 23*(1–2), 315–324.

Dale, B., Bay-Larsen, I., & Skorstad, B. (Eds.). (2018). *The will to drill-mining in Arctic communities*. Springer International Publishing.

Easterly, W. (2006). Reliving the 1950s: The big push, poverty traps, and takeoffs in economic development. *Journal of Economic Growth, 11*(4), 289–318.

Fleischer, J. (2000). *Grønlands Lumumba*. Atuagkat.

Hamilton, L. C., Brown, B. C., & Rasmussen, R. O. (2003). West Greenland's cod-to-shrimp transition: Local dimensions of climatic change. *Arctic, 56*(3) 271–282.

Heinrich, J. (2017). Independence through international affairs: How foreign relations shaped Greenlandic identity before 1979. In K. S. Kristensen & J. Rahbek-Clemmensen (Eds.), *Greenland and the international politics of a changing arctic* (pp. 28–37). Routledge.

Holm, P. (2012). World war II and the "great acceleration" of North Atlantic fisheries. *Global Environment, 5*(10), 66–91.

Jonsson, I. (1996). Reflexive modernization, organizational dependency and global systems of embedded development—a post-colonial view. *Cultural and Social Research in Greenland, 95*(96), 130–145.

Humphreys, D. (2019). The mining industry after the boom. *Mineral Economics, 32*(2), 145–151.

Kapel, F. O., & Petersen, R. (1986). Subsistence Hunting--the Greenland Case. In: Donovan, G.P., ed. Aboriginal/subsistence whaling. Cambridge: IWC, Special Issue No. 4.

Kleivan, H. (1969). Dominans og kontrol i moderniseringen af Grønland. *Grønland i fokus, Copenhagen; Nationalmuseet* (pp. 141–166).

Kleivan, H. (1984). Contemporary Greenlanders. *Handbook of North American Indians, 5*, 700–717.

Langgård, K.. (1999). "Fishermen are weaklings" perceptions of fishermen in Atuagalliutit. In O. Marquardt, P. Holm, & D. Starkey (Eds.), *From sealing to fishing: Social and economic change in Greenland, 1850–1940 Studia Atlantica* (4th Ed.). Fiskeri og Søfartsmuseet.

Larsen, F. B. (1992). The quiet life of a revolution: Greenlandic Home Rule 1979–1992. *Études/Inuit/Studies, 16*(1), 199–226.

Marquardt, O. F. (1996). The employees of the Royal Greenland Trade Department (1850–1880). *Études/Inuit/Studies, 20*(1), 87–112.

Marquardt, O. F. (1999a). An introduction to colonial Greenland's economic history. In O. Marquardt, P. Holm, & D. Starkey (Eds.), *From sealing to fishing: Social and economic change in Greenland, 1850–1940 (Studia Atlantica)* (4th Ed.). Fiskeri og Søfartsmuseet.

Marquardt, O. F. (1999b). A critique of the common interpretation of the great socio-economic crisis in Greenland 1850–1880: The case of Nuuk and Qeqertarsuatsiaat. *Études/Inuit/Studies, 32*(1–2), 9–34.

Marquardt, O. F., & Caulfield, R. A. (1996). Development of West Greenlandic markets for country foods since the 18th century. *Arctic, 49*(2), 107–119.

Ministeriet for Grønland. (1981). *Grønland 1981*. Årsberetning.

Nuttall, M. (2005). *Encyclopedia of the Arctic*. Routledge.

Petersen, H. C. (1990). *Grønlændernes historie - fra urtiden til 1925*. Atuakkiorfik.

Petersen, R. (1989). Traditional and contemporary distribution channels in subsistence hunting in Greenland. In J. Dahl. *Keynote speeches from the Sixth Inuit Studies Conference Copenhagen* (pp. 81–89). Institut for Eskimologi, Københavns Universitet.

Petersen, R. (1995). Colonialism as seen from a former colonized area. *Arctic Anthropology, 32*(2), 118–126.

Petterson, C. (2012). Colonialism, racism and exceptionalism. In K. Loftsdóttir & L. Jensen (Eds.), *Whiteness and postcolonialism in the Nordic region* (pp. 41–54). Routledge.

Poppel, B. (2005). G-50. In M. Nuttall (Ed.), *Encyclopedia of the Arctic* (Vol. 1). Routledge.

Poppel, B. (2006). Interdependency of subsistence and market economies in the Arctic. In S. Glomsrød & I Aslaksen (Eds.), *The economy of the North* (65–80). Statistics Norway.

Poppel, B., & Stenbæk, M. (2005). Birthplace criteria. In M. Nuttall (Ed.), *Encyclopaedia of the Arctic* (Vol. 1, pp. 261–262). Routledge.

Rasmussen, R. O. (2000). Formal economy, renewable resources and structural change in West Greenland. *Études/Inuit/Studies*, 24 (1): 41–78.

Rasmussen, R. O., & Hamilton, L. C. (2001). *The development of fisheries in Greenland with a focus on Paamiut/Frederikshåb and Sisimiut/Holsteinsborg.* Roskilde University.

Rud, S. (2017). *Colonialism in Greenland: Tradition, governance and legacy.* Springer.

Seiding, I. (2013). *"Married to the Daughters of the Country": Intermarriage and intimacy in Northwest Greenland ca. 1750 to 1850* [Unpublished doctoral dissertation]. *Grønlands Universitet.*

Sørensen, A. K. (2007). *Denmark-Greenland in the twentieth century* (Vol. 341). Museum Tusculanum Press.

Tejsen, A. V. S. (1977). The history of the Royal Greenland Trade Department. *Polar Record, 18*(116), 451–474.

4 The Greenlandic employment system

Thorny tasks and ambitious targets

Helle Holt and Frederik Thuesen

Abstract

The Greenlandic employment system aims to facilitate good matches between jobseekers and employers, given a labour market characterized by challenging structures in terms of geography, available job types, and skill distribution. In 2015, the Parliament of Greenland enacted legislation to reform this system. Effective 1 January 2016, this reform led across Greenland to the establishment of 17 Majoriaq centres (i.e., job centres), where citizens of working age must apply for public benefits in case of unemployment. The Majoriaq centres are responsible for casework, counselling, and skills upgrading related to both unemployed youth and adults. Each Majoriaq has three sections – a labour market section, an education counselling section, and a skills upgrading section. This chapter analyses the Greenlandic employment system based on qualitative and quantitative data collected in 2019. The results focus on the structural labour market conditions in which the different Majoriaq have to operate, the internal organization, caseworker resources, the tools, and targets the caseworkers must implement as well as the external collaboration with the municipal social services departments and with private companies. Finally, the chapter discusses some challenges Greenland faces in obtaining a well-functioning employment system and labour market.

4.1 Introduction

Greenland's labour market is a vast archipelago of local labour markets separated by ice, snow, water, mountains, and enormous distances (see Chapter 1). On the demand side of the labour market, the types of employment available span traditional blue-collar jobs within fishing and mining as well as specialized occupations in construction and tourism. Some of these occupations require limited skills, but a growing proportion of the jobs available require higher education and more qualifications in enterprises operating in a competitive global market. On the supply side of the labour market, a relatively large proportion of the working population have no more education than nine years of schooling from the Greenlandic primary school. Many fail to take the final exam in primary school and later need to take remedial courses to obtain the primary school diploma (Pedersen et al., 2018). A substantial share of Greenland's citizens of working age between 16 and 65 years are outside the labour market and living on social benefits. The Greenlandic employment system aims to facilitate good matches between jobseekers and employers, given a labour market characterized by challenging structures in terms of geography, available job types, and skill distribution. An efficient employment system is required to contribute to economic growth and development in Greenland.

In 2015, the Parliament of Greenland enacted legislation to reform the employment system (Inatsisartut-law no. 28, December 2015 on Job-, Counselling-, and Skills Upgrading Centres). Effective 1 January 2016, the reform led across Greenland to the establishment of 17 Majoriaq centres (i.e., job centres), where unemployed people can go for advice and information about available jobs. The primary mission of the new Majoriaq centres is to provide a bridge between the educational system, the labour market, and the private sector. Previously, local labour market offices that served the unemployed existed in the main towns. Those labour market offices were separate from the municipal Piereersarfiit centres. Piereersarfiit existed to provide educational counselling and courses targeting youth lacking the public school diploma so they could prepare for this exam (Boolsen, 2012). The 2016 reform merged the local labour market offices and the Piereersarfiit into the new Majoriaq centres, which were then assigned responsibility for casework, counselling, and skills upgrading relating to both unemployed youth and adults. Hence, each Majoriaq has three sections – a labour market section, an education counselling section, and a skills upgrading section (see Figure 4.3).

The reform left the municipal social services offices unaffected. Those offices are responsible for casework in the field of benefit recipients in need of public social assistance because of substance abuse, child neglect, disabilities, etc. The social services offices are also responsible for paying out different types of benefits for different types of unemployed benefit claimants – primarily unemployment benefits (arbejdsmarkedsydelse) and public benefits (offentlig hjælp). Hence, unlike the (former) British Jobcentre Plus agencies that brought together the services of a benefits agency and an employment service (Karagiannaki, 2007), benefits and employment services in Greenland are split between the municipal social services offices and the Majoriaq, respectively. Although the Majoriaq and the social services offices have separate missions, they need to collaborate to enhance the employment prospects of the most vulnerable benefit recipients and to implement interventions targeting these benefit recipients ingrained in the 2016 reform. Previous research, including studies focusing on municipal Danish Jobcentres and the municipal Danish social services departments (Kjeldsen, 2020), indicates that such collaboration can be difficult (Askim et al., 2011; Bredgaard, 2011; Knuth & Larsen, 2010; Larsen et al., 2020). Below we will analyse whether this is also the case in Greenland.

This chapter analyses the Greenlandic employment system based on qualitative and quantitative data collected in 2019. Below the management level, our data collection targeted employees in the labour market sections of the Majoriaq. In the next section, we describe our data and methods. Second, we outline the structural labour market conditions in which the different Majoriaq have to operate. Third, we describe the main categories of the social security system, and fourth we describe the purpose of the Greenlandic employment system and the organizational structure of the Majoriaq. Fifth, we analyse the interventions and tools the caseworkers must implement and the tools they dispose of, and sixth we analyse staffing, qualifications, and relations to the local community. Seventh, we analyse the collaboration between the Majoriaq and the municipal social services departments, and eighth we analyse the collaboration between Majoriaq and private companies. Finally, we draw some conclusions and discus some of the challenges Greenland faces in obtaining a well-functioning employment system and labour market.

4.2 Data and methods

The analyses draw on three different sources of data collected and analysed in 2019 (Holt et al., 2019). The Danish Center for Social Science Research (VIVE) collected the data as part of an evaluation of the 2016 reform of the Greenlandic employment system commissioned by the government of Greenland (Holt et al., 2019). First, during spring 2019, we conducted interviews in six different towns with four different groups of respondents: (a) the municipal employment manager, (b) the Majoriaq manager, (c) employees in Majoriaq labour market section, and (d) unemployed benefit recipients who had received assistance from the Majoriaq. In total, we conducted 41 interviews. In the following analyses, we will only use interview data from the first three groups of respondents. Second, during summer 2019, we conducted a web-based survey among the Majoriaq's 17 managers (response rate 94%), and another web-based survey among 74 employees in the Majoriaq labour market sections. Among the employees, we obtained 58 full replies, equivalent to a response rate of 78%. Both managers and employees had access to a Greenlandic and a Danish version of their respective questionnaires. Third, we gained access to administrative data from Statistics Greenland concerning the labour market as well as data from the Ministry of Natural Resources and Employment concerning activities in the Majoriaq. The most recent year with available full administrative data was 2017. Concerning methods, for qualitative data we extract meaning and the main tendencies from our interviews (Kvale, 1996); for quantitative data, we mainly present descriptive statistics (Agresti, 2018).

4.3 The Greenlandic labour market – centre and periphery

In 2017, 37,877 individuals were of working age (15–65 years) in Greenland out of a total population of 55,800 persons (Holt et al., 2019). The average employed population consisted of 27,250 people. Hence, 10,627 individuals of working age, equivalent to 28% of the working age population, were unemployed or permanently outside the workforce on early retirement pensions. Hence, in 2017, the employment rate in Greenland was 72%. For a comparison, in 2017, the Danish employment rate was 73%, which is highly equivalent to the rate in Greenland (see www.statistikbanken.dk). According to Statistics Greenland, the unemployment rate among permanent residents age 18–65 years was 6.8% in 2017, and showing a decreasing tendency from 10.3% in 2013.

A relatively low level of education characterizes Greenland's workforce. Approximately half of the employed have only the public school diploma (i.e., 9 years of schooling), while roughly 30% have vocational training. The rest has either an upper secondary school diploma or tertiary education (see also Figure 4.1). In the other Nordic countries, around 25% have no education above the lower-secondary level (Statistics Greenland, 2020).

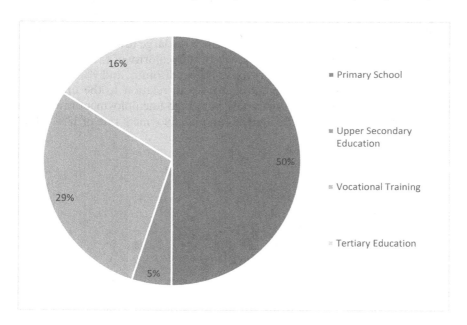

Figure 4.1 Levels of education among employed persons in Greenland, 2017.

Note: N = 25.972, own calculations based on data from Statistics Greenland. See also the distribution of educational attainment among all persons in Greenland 15–64 years, 2018 calculated by Statistics Greenland (2020, p. 23) that shows a very similar picture although the share of persons with no more education than primary school is 5 pct.-point larger when taking the full population. Hence, the employed are slightly better educated than the population at large.

The employment rate in Greenland is highly seasonal since fishing and the service industry (e.g., tourism) represent two dominant trades. Therefore, the unemployment rate is higher during the winter months than during the summer. In 2017, 23,832 people held a job in January, while this figure had increased to 26,434 in June (Holt et al., 2019, p. 20). Public administration (i.e., people working for the Greenlandic government or the municipalities) accounts for 40% of the employed. Public administration is hence the trade sector that employs the highest share of people. Fishing, hunting, and agriculture comes second, employing 17% of those employed. Wholesale, retail trade, and repairs, as well as transportation, mail, and telecommunication are third and fourth, respectively. In some of these sectors (e.g., services and mining), foreign workers are common. Hence, a number of foreign workers from outside of Greenland, the Faroe Islands, and Denmark obtain permission to work in Greenland every year. In 2017, the Danish Agency for Recruitment and Integration granted 220 of these permissions (according to the Danish agency itself).

Greenland has five municipalities and 17 main towns that each has a Majoriaq. The capital Nuuk is by far the largest town with 18,326 inhabitants, Sisimiut is second largest with 5,582 inhabitants, and Ilulissat is third largest with 4,670 inhabitants (Statistics Greenland, 2020, p. 4). The jobs distribution reflects these demographic patterns since the large towns such as Nuuk, Sisimiut, and Ilulissat are the hometowns of the majority of those employed. Of all those employed, 36% live in Nuuk, 12% in Sisimiut, and 9% in Ilulissat (see Figure 4.2). The reverse picture emerges in relation to the unemployment rate where some small towns have very high unemployment rates. The secluded east coast town of Tasiilaq has an unemployment rate of 18%, while

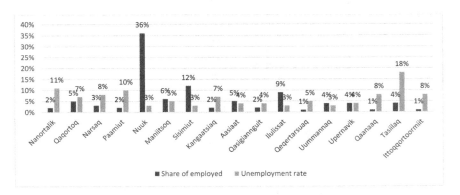

Figure 4.2 Distribution of employed persons[1] and unemployment rate[2] in the 17 Majoriaq towns.

Note: 1) Distribution of share of employed persons across the 17 main towns of Greenland: N = 25.972 (percentages sum to 100 pct.). 2) Unemployment rate: N = 37.877. Remark that the unemployment rate is calculated as full-time unemployed persons as a share of the working age population of each town (and therefore the percentages do not add up to 100 pct.).

in Nuuk this rate is only 3% (see Figure 4.2 and Holt et al., 2019, p. 24). Hence, large regional differences characterize Greenland's unemployment rate. Moreover, the fishing industry is a dominant trade, and the labour market is highly dependent on this specific trade. Nuuk is an exception to this rule as it is home to the Greenlandic government and central administration. In Nuuk, public administration employs a large share of the working population.

The three largest Majoriaq centres, Nuuk, Sisimiut, and Ilulissat, all serve expanding local economies – something that the different unemployment levels in Figure 4.2 corroborate. In these three towns, local economies generate a labour demand that their respective Majoriaq may not always be able to fulfil. Economic growth and occasional labour shortages is the backdrop for local Majoriaq managers and employees in those towns who, in our interviews, recount relatively good opportunities for benefit recipients with health problems or no education to find jobs. Likewise, in the bigger towns, caseworkers find it easier to convince vulnerable unemployed benefit recipients that they should engage themselves in job clarification processes and disablement rehabilitation jobs because such engagement may ultimately lead to formal employment.

In contrast, many smaller towns located remotely from these expanding local economics tend to experience economic stagnation and a higher level of unemployment – as equally revealed in Figure 4.2. Therefore, in those towns the Majoriaq employees are less optimistic in terms of the employment prospects for the least readily employable benefit recipients. In those towns, even the job-ready unemployed (in the so-called "Match Group 1", see below) face a very limited number of job openings, and the prospects for employment for vulnerable unemployed benefit recipients are even bleaker. Likewise, companies are less willing to participate in work capacity tests and establish disablement rehabilitation jobs.

4.4 Social security

Greenland's social security system follows the Nordic welfare model (Statistics Greenland, 2020, p. 15). This system secures the income of working age individuals who do not have salaried employment. The Greenlandic social security system has the following main types of transfer income:

- *Unemployment benefits* is a public transfer income available for 13 weeks in case of unemployment. Access requires a previous job – more specifically, 182 hours of salaried work throughout the previous 13 weeks.[1] After 13 weeks, recipients may apply for public benefits.
- *Public benefits* is a public transfer income available to unemployed individuals if all other sources of income are exhausted. Public benefits are equivalent to the Danish cash benefits scheme that unemployed persons may apply for if they have no other sources of income and private means below a certain (low) threshold (Pedersen et al., 2018; see also Chapter 6 in this book).[2]

- *Early retirement pension* is a public transfer income to individuals of working age affected by reduced working capacity. The benefit has three levels – lowest, intermediate, and highest level – dependent on the applicants' residual work capacity. Hence, access to one of those levels of early retirement pension requires a public assessment of work capacity. Early retirement pension used to be a permanent benefit type once awarded. However, a reform of the disability system that took effect on 1 July 2016, changed this pension into a temporary benefit, and Majoriaq now has to assist the social services departments in reassessing the recipients' work capacity every five years. Moreover, the reform implied that a large share of existing recipients of early retirement pension should have their cases reassessed within five years from 1 July 2016. This requirement has put both the Majoriaq and the municipal social services departments under pressure to meet this deadline.[3]

- *Flexi-jobs, disease benefits, and maternity leave* are types of transfer income with fewer recipients than the previous three types. However, the flexi-job scheme is important because of the Greenlandic political aspirations that more benefit recipients – who would previously be awarded early retirement pension – take up a flexi-job that is less expensive in terms of public funding. A flexi-job is a type of publicly subsidized job in a private or public workplace with working hours and a workload adapted to persons with permanent work capacity reductions. A similar scheme has existed in Denmark since 1998, where the scheme during recent years has been a success in terms of providing access to the labour market for people with a permanently reduced work capacity (see also Barr et al., 2019).[4] In Greenland, the scheme has existed since 2001; however, so far, it has not been equivalently successful in terms of providing salaried work for people who would otherwise receive early retirement pension. The share of people on early retirement has been stable over the years (Statistics Greenland, 2020, p. 15).

In 2017, as a monthly average 697 individuals received the unemployment benefits, equivalent to 2% of the full working-age population. As a monthly average, 2,085 individuals received the early retirement pension, while 2,654 people received the public benefits, equivalent to 6% and 7% of the working-age population, respectively (Holt et al., 2019, p. 24). Hence, compared to the two other types of public benefits, the share of individuals receiving unemployment benefits is low.[5]

4.5 The purpose of the Greenlandic employment system and the Majoriaq

1 January 2016, the new Law on Job-, Counselling-, and Skills Upgrading Centres (Majoriaq centres) entered into force. There are 17 Majoriaq in Greenland's 17 major towns across the five municipalities of Greenland.

Figure 4.3 The three sections of the Majoriaq.

Hence, every municipality has several Majoriaq centres and each Majoriaq centre has a number of settlements to serve. Each municipality has one employment manager who is overall responsible for securely implementing legislation and policies in the employment field in the municipality and its Majoriaq centres. The 17 Majoriaq are crucial institutions in terms of implementing Greenlandic employment policies. A Majoriaq has three overall sections and types of tasks (see Figure 4.3).

1 The counselling section focuses on counselling unemployed benefit recipients in the fields of education, continuing education, and training.
2 The skills upgrading section conducts courses, mainly the primary school diploma course ("PSD course", in Danish "FA-kursus", i.e., Primary School Diploma) targeting youth or adults who need to complete their primary school and obtain the diploma or improve their marks.
3 The labour market section focuses on matching unemployed benefit recipients with available jobs. The labour market section is also responsible for initiating and conducting the job clarification process as well as the Counselling and Motivation course (CM-course, see below) to improve the employment prospects of those vulnerable benefit recipients who may still have some work capacity.

In this chapter, we only focus on the labour market section and its targets and tasks.

In terms of organization, each Majoriaq has a manager and a number of employees. The labour market section typically consists of two labour market consultants and two clerical workers operating the front desk. These two clerical workers are, inter alia, responsible for stamping the unemployment form of those receiving the unemployment benefits who need to register as unemployed in the Majoriaq every second week. According to our survey to the Majoriaq managers, the Majoriaq have on average 12 employees each, including all three sections (minimum 3 and maximum 42). The labour market sections have on average four employees each (minimum 1 and maximum 15). Hence, the total size of a Majoriaq and its labour market section varies substantially depending on whether the Majoriaq is located in a major or a minor town.

Employees in the labour market section primarily assist unemployed benefit recipients in finding a job. As part of this assistance, they evaluate whether a benefit recipient needs skills upgrading or other targeted interventions and draw up an action plan. They also categorize unemployed benefit recipients into one of three match groups:

- *Match group 1 – the labour market ready.* This group of benefit recipients is often unemployed because of seasonal layoffs, or potentially because a company has closed. If seasonally unemployed, these benefit recipients need to get their stamp and wait for, for example, job offers in fishing or tourism to show up again. While unemployed, the Majoriaq may offer these benefit recipients continuing training or education to improve their employability (the so-called Project Competencies Development – PCD-courses).
- *Match group 2 – the activity ready.* These benefit recipients are not immediately employable but may come closer to employment via different types of interventions, typically a rehabilitation job. A job clarification process often precedes those interventions (see below).
- *Match group 3 – the not activity ready.* These benefit recipients are, typically because of major health problems or permanent disability, not employable and therefore not activity ready. Casework in relation to this group often focuses on their application for early retirement pension or on a reassessment of their entitlement for such pension. Still, benefit recipients in this group often are required to pass through a job clarification process and, dependent on the outcomes of the work capacity test, some subsequently take up a rehabilitation job.

In terms of concrete work hours spent on each of those groups, our interviews with Majoriaq employees revealed that they spent little time on Match Group 1, and much more time on benefit recipients from Match Groups 2 and 3. Unemployed benefit recipients from the latter two groups are very diverse (Antropologerne, 2016). Many have no education beyond primary school as well as problems relating to substance abuse, social and familial issues, and physical and mental health problems – providing the caseworkers with a difficult task of furthering their progress towards employment. A substantial number of these benefit recipients have never had a job, or perhaps they had one many years ago. If they are required to undergo a reassessment of their entitlement to early retirement pension, still they may not want a job but rather just want to keep their early retirement pension. However, the recent Greenlandic labour reforms aim at strengthening the Majoriaq's ability to guide more of these benefit recipients into employment to save public funds and provide the private sector with a greater pool of labour supply. Still, the Majoriaq's ability to realize this goal and implement the concomitant legislation faces a number of challenges, which we will outline below.

4.6 Interventions targeting vulnerable unemployed benefit recipients

Increasing the employment rate of benefit recipients in match group 2 and – if possible – match group 3 is a central target of recent labour market reforms in Greenland. The caseworkers of the Majoriaq make use of three essential tools to achieve this target: the job clarification process, the counselling and motivation course (that may be an element in the job clarification process), and the subsidized disablement rehabilitation jobs (see Figure 4.4). Given their centrality in the Greenlandic employment system, we will outline below these tools' content and practical functioning.

If job readiness and work capacity of an unemployed citizen is in doubt, Majoriaq caseworkers may initiate a job clarification process. This process has a minimum of two components: the work capacity assessment (or test) and the subsequent tripartite talks of the assessment's outcomes. The assessment takes place in a public or private workplace and lasts a minimum of two weeks. The assessment typically precedes a disablement rehabilitation job but may also be an element in processing early retirement pension applications and re-evaluations of an entitlement to such pension. Succeeding the assessment, the Majoriaq sets up a tripartite meeting with the unemployed citizen, the Majoriaq caseworker, and an employer to evaluate the test's outcomes. After this meeting, the Majoriaq caseworker appraises the case's facts and takes a decision on what should happen next – for example, if the citizen is ready for a disablement rehabilitation job or whether the caseworker should recommend her/ him for an early retirement pension. Often, this appraisal and decision is also a point on the agenda of a collaboration council meeting (see below).

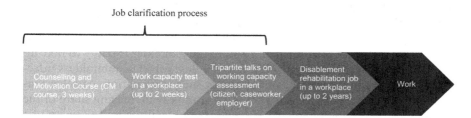

Figure 4.4 Interventions in an ideal sequence guiding vulnerable unemployed benefit recipients into work.

4.6.1 The counselling and motivation (CM) course

A CM course may precede the work capacity test as part of the job clarification process if the caseworker estimates that the unemployed citizen has limited work experience, a lack of work motivation, and/or a need for guidance on how to handle a job. The course lasts four hours a day for three weeks and includes follow-up meetings up to a year after (Holt et al., 2019, pp. 70–72; see also Broberg, 2019). The Ministry of Labour has developed a course manual, and has trained the Majoriaq caseworkers in how to conduct the course. The material for the course consists of, for example, different presentations, YouTube films, and guidelines for how to motivate the participants for actively seeking employment (Broberg, 2019). Mostly, the Majoriaq managers and caseworkers that we interviewed found this course highly relevant in relation to the target group. One caseworker at a large Majoriaq who had conducted several CM courses said, "My estimate is that these courses are a great success, and the material to be used [in relation to the course] is of high quality and works. It's my bible".

Nevertheless, the interviewed managers and caseworkers also found that the courses imposed considerable psychological strain on both caseworkers and participating benefit recipients. During the CM course, the caseworkers work with benefit recipients who often have different types of mental and physical health problems. One interviewed caseworker said that during a recent course, four participants had shared experiences of massive neglect of care during their childhood. Caseworkers typically lack appropriate education and tools to handle such experiences that may also occupy the minds of other course participants and blur the course's job focus.

4.6.2 Disablement rehabilitation job

Building upon the job clarification process, a disablement rehabilitation job is the tool caseworkers typically use seeking to provide a concrete path into the labour market for vulnerable benefit recipients. Both public and private workplaces can be the setting of such a job that may last up to two years with periodical follow-up meetings in the Majoriaq. The purpose of a disablement rehabilitation job is to develop the jobholder's work capacity to a point where a company may be willing to hire this person. Nevertheless, often the contract between the jobholder and the workplace ends prematurely, that is, before scheduled and before the jobholder is ready to take up an ordinary job (part time or full time) or a flexi-job. The jobholders infrequent work attendance can be one cause for such termination. Another reason that workplaces are required to pay a gradually increasing share of the jobholder's salary, and many workplaces are only willing to do this up to a certain point. One caseworker from a small Majoriaq said, "Unfortunately, often the disablement rehabilitation contracts are broken ahead of time.

Either by the citizen, because some private problems come up, or by the company because it costs money to have this employee".

Still, the disablement rehabilitation jobs are the most important practical tool that caseworkers can use to provide a path into the labour market for benefit recipients with limited work capacity.

4.7 Staffing, qualifications, and local community

Externally, stagnating local economies in some towns can be a challenge to successfully deploying the tools that employees in the Majoriaq use. Internally, both managers and employees quote lack of employee resources and qualifications as important obstacles to implementing the targets of recent labour market reforms. Hence, both managers and employees find the Majoriaq's labour market sections understaffed and underqualified in terms of implementing complex labour market legislation and dealing adequately with the personal problems, health issues, etc. that often characterize unemployed benefit recipients belonging to Match Groups 2 and 3.

For Majoriaq managers, our survey indicated that they had rather diverse levels of education and leadership training. In terms of education, these levels spanned from clerical worker education to academic university degrees. Most managers had vocational training or tertiary education within the education and counselling fields. Several managers had taken leadership courses. The survey to the Majoriaq employees showed that the main part of employees in the labour market section had vocational training, typically as commercial and clerical workers. Very few had some sort of tertiary education, for example, as social workers or as public employment consultants. Some employees had previously worked as specialist teachers in the previous Piareersarfiit centres. Some had received training on how to conduct the CM courses facing vulnerable benefit recipients. Still, viewed in light of the complexity of the tasks the employees have to handle, the level of education among both managers and especially employees is low.

In our interviews, the municipal employment managers, the employees themselves, and the Majoriaq managers generally agreed that employees lacked sufficient qualifications to take care of the complex tasks located within the Majoriaq labour market sections. Often, the caseworkers have to confront problems that they are neither educated nor sufficiently trained to handle – this being the case regardless of whether these employees work in a large or a small Majoriaq. Every four years, the Danish National Institute of Public Health conducts a public health survey in Greenland. To illustrate the depth of problems that may characterize vulnerable unemployed benefit recipients – problems that employees in the Majoriaq often have to confront – the 2018 report showed that among persons born in 1995 or later, 20% had been subjected to sexual abuse during their childhood (before the age of 18 years). For people born from 1970–1979, this is the case for 40% (Larsen et al., 2019, p.18). Violence against children and alcohol abuse among their parents has shown a decreasing tendency during recent years. Still, among children born in 1995 or later, around one-third grows up in a home characterized by either violence or alcohol abuse (Larsen et al., 2019, pp. 16–17).

One employee said, "Our competencies are off balance in relation to the work tasks we have to carry out".

In a long-term perspective, managers and employees called for more caseworkers with formal social worker education and for stronger legal and psychological competencies within the Majoriaq. In a short-term perspective, employees called for more supervision. Concerning both complex legislation and unemployed benefit recipients with psychological problems, one employee from a small Majoriaq said:

> We find it difficult to follow the very complex legislation that should be the basis of our work. We have tried to read this legislation in both Danish and Greenlandic several times. Nor are we educated as social workers or psychologists but this [such education] would have be good because the unemployed have many questions.

Several managers shared the view that the employees lacked sufficient competencies. One Majoriaq manager said:

> As a manager, I think that the employees have inadequate competencies. It takes strong people to confront heavy problems and the employees are not necessarily such people. We lack some tools to handle those difficult psychological issues the employees have to deal with. We lack supervision, and we lack social worker competencies in our daily life.

One municipal employment manager believed that both Majoriaq managers and employees lacked adequate qualification:

> Lack of competencies among both Majoriaq employees and managers is a great challenge. We simply lack social worker competencies. Employees and middle managers are typically educated as clerical workers. They are skilled when it comes to routinized administrative work, but not in relation to those complex tasks they may in reality have to face in a Majoriaq.

Our surveys to managers and employees also shed light on the employees' competencies. In both surveys, we asked the respondents whether employees in the employment section had adequate competencies to perform a number of work tasks: (a) counselling in relation to job search, (b) counselling in relation to education, (c) initiating disablement rehabilitation jobs in companies, and (d) conducting a CM course. Approximately half of both groups of respondents indicated on a 5-point scale that the employees merely had "very limited", "limited", or "some" degree of competencies to perform these tasks. The lowest estimates pertained to employee competencies in terms of teaching CM courses. Of the managers and employees interviewed, 38% and 31%, respectively, replied that to a very limited or a limited degree employees had sufficient competencies to run these types of courses.

Implementation research indicates that street-level bureaucrats' skills and motivation are important factors to successfully implementing policy goals and new legislation (Winter, 2014). Therefore, recruiting and retaining competent employees is crucial for the Majoriaq's ability to carry out complicated unemployment-related casework in a successful way. For both recruiting and retaining employees, our interviews indicated that some Majoriaqs faced difficulties and that several work environmental issues caused these difficulties, including a stressful everyday work life, mediocre pay, and a job with a bad reputation. Concerning these issues, two interviewees (an employee and a manager, respectively) said:

> Here it is only the fittest who survive, and therefore it [the Majoriaq centre] has become a transition camp. We have a very bad reputation.
>
> Two employees have quit for other jobs in the municipality where the salary is better, the workload is lighter and the workday is less intensive.

Nonetheless, data from our two surveys paint a somewhat less bleak picture of the Majoriaq's recruitment situation. Among the managers, 73% of our respondents replied on a 5-point scale that it is "very rare" or "rare" that their employees find a job in another workplace. Descriptive statistics from our employee survey showed an average tenure among employees of four years in their current job. Still, these figures do not rule out the possibility that some Majoriaq may be a "transition camp" – or that some long term "stable" employees may hold such an experience.

Another challenge to the casework of many Majoriaq employees relates to working and living in small communities "where everybody knows everybody." Benefit recipients who reveal personal problems during a meeting in the Majoriaq may live next door to the caseworker and their family. Such proximity can be a psychological burden to the caseworker and even to a higher extent if the law requires a caseworker to sanction an unemployed citizen who, for example, fails to turn up for a meeting. Tensions may also arise if a caseworker and their colleagues decide that a "neighbour" needs to undergo a reassessment of their entitlement for early retirement pension and/or based on such reassessment take away this entitlement. One employee in a small Majoriaq said, "It can be difficult to take away the money from people when you know the family and also the consequences". Another employee said, "It is hard to sanction in a small society. Once, I was verbally attacked because I had sanctioned a citizen".

In small communities, proximity and traditional tightly knit reciprocity relations imply that implementing legislation can come at a psychological cost to the local caseworkers. American sociologist Alexandro Portes (1998, pp. 16–17) argues "negative social capital" characterize tightly knit communities that restrain individual freedom. Clearly, such community relations can also impede implementing universalistic state norms if they run counter to the interests of some community members (see also Smylie & Evans, 2006).

Overall, both interview and survey data show that weak employee competencies to work with unemployed benefit recipients with complex personal problems impede implementing recent labour market policy goals. Moreover, implementing strict employment legislation in small communities where caseworkers and clients know each other well and live next door can be difficult.

4.8 Collaboration with the municipal social services department

Two types of collaboration are crucial to the Majoriaq labour market sections' ability to realize the goal of bringing unemployed benefit recipients into work. In this section, we analyse the Majoriaq's collaboration with the municipal social services departments that is important in relation to dealing with hard-to-place unemployed benefit recipients. In the next section, we analyse the Majoriaq's collaboration with local employers in public and private workplaces that is important in relation to all types of unemployed benefit recipients. In Holt et al. (2019, pp. 53–55), we have also analysed the functioning of internal collaboration within Majoriaq (i.e., between the three different sections), which is also important to the labour markets sections' ability to realize its goals. Generally, this internal collaboration appears to function relatively well. Therefore, we here choose to focus on the Majoriaq's external collaborative relations that both managers and employees often find wanting.

In contrast to the internal collaboration within the Majoriaq, collaboration with the municipal social services departments is often troublesome – despite this relationship being important because both departments assist some of the same benefit recipients. More specifically, Majoriaq typically collaborates with the social services departments concerning early retirement applications or early retirement pension reassessments. According to several interviewed Majoriaq managers and employees, collaboration with the social services department is important, but difficult. One of these managers said:

> In relation to the reassessments of early retirement pension entitlements, the social services departments work out the social and health profiles of the applicants, while we conduct the work capacity test, and, if relevant, help find a disablement rehabilitation job. However, we cannot work with those citizens that the social services departments send us—citizens with health problems and citizens who cannot work.

The quote above is an example of common perceptions among some Majoriaq managers who often perceive the benefit recipients in the target group of a potential early retirement pension as too weak to possess a realistic labour market perspective. Despite many expressions of frustration in our interviews concerning the relation between Majoriaq and the social services department, our surveys show a more positive picture. In surveys to both managers and employees, we asked the respondents to rate the quality of collaboration between Majoriaq and the social services departments pertaining to long-term sick listed benefit recipients, benefit recipients undergoing disablement rehabilitation, and applicants for early retirement pension. For the managers, 50% rated the quality as "fair", and 50% as "good" or "very good"; hence, remarkably, no manager respondent replied "very bad" or "bad." Generally, managers were more sceptical of the quality of collaboration with the health sector pertaining to these citizen groups than of the quality of collaboration with the social services department. Compared to the managers, the employees viewed the quality of collaboration with the social service department with more scepticism. Thus 27% of the employee respondents rated collaboration as "very bad" or "bad", 34% as "fair", and 39% as "good" or "very good." Employee ratings of collaboration with the health system was similarly sceptical to the managers' ratings.

Formally, the so-called council of collaboration provides the legally instituted framework for collaboration between Majoriaq and the municipal services departments. In principle, meetings in the council take place at least once a month and involve representatives from Majoriaq, the social services department, and if relevant, local schools or the health system. The meetings aim to handle difficult cases, for example, where employees from Majoriaq and/or the social services department need to discuss which match group a citizen belongs to or the right action plan for a particular citizen. Ideally, the council should also be able to handle cases swiftly so that benefit recipients need not experience unnecessary waiting time. In some towns, managers estimated that the council functioned according to the intentions. One manager said, "The council of collaboration works fine. We hold the meetings once a week, but only if there is a case. As the Majoriaq manager, I chair the meeting, and we deal with matters of dispute".

Among the interviewed employees, some had participated in meetings in the council, while others barely knew what was going in the council. In general, the interviewed employees found the council to function in a suboptimal manner and unable to provide the intended framework for handling difficult cases. One employee and one manager, respectively, said:

The council of collaboration does not function. The social services department is to slow in terms of processing their cases, and they do not show up at the meetings.

We have not yet been successful in terms of making the council of collaboration function properly. We find that the social services department

lets us down by not sending a representative or not the same representative. I mean, it requires some continuity in the group of participants so
that we may establish some sort of meeting culture.

Our survey to the managers contained a question concerning the meeting
frequency of the collaboration council. Among the respondents, 75% indicated that meetings take place once or twice a month; the rest indicated
either a higher or a lower meeting frequency. Overall, our interviews and
survey data indicate that Majoriaq and the social services departments
do in fact collaborate on assessing the work capacity of vulnerable unemployed benefit recipients, on testing this capacity, and on making plans
and taking decisions relevant to rehabilitating such benefit recipients.
However, according to our Majoriaq interviewees and respondents, the
quality of this working relationship may improve – which appears all the
more important given the skills deficit described in the previous section
and the complexity of problems that characterize the benefit recipients in
question.

4.9 Collaboration with the local workplaces

The main purpose of the Majoriaq labour market sections is to facilitate
the matching process linking unemployed jobseekers with available jobs.
Therefore, collaboration with local private enterprises and public workplaces is also crucial. However, as described in relation to Figure 4.2, very
different conditions characterize the local labour markets, depending on
whether the local economy is stagnant or expanding, and these conditions
provide different opportunities for collaboration. Moreover, managers and
employees approach collaboration with local workplaces differently from
town to town. Nonetheless, two main challenges in most towns are (a) to
incite the local enterprises and other workplaces to take advantage of the
local labour supply, and (b) to take part in the rehabilitation and development of the working capacity of vulnerable workers.

4.9.1 Stimulating companies to take advantage
of the local labour supply

Inciting local private enterprises to hire local unemployed workers is a challenge that affects all match groups—not just Match Group 2 and, potentially, 3 – but also Match Group 1. As to Match Group 1, our interviews
revealed that the employees in the Majoriaq approached the enterprises
in two ways. We will name these the *offensive* and the *defensive* strategies.
Below we quote an employee from a Majoriaq in one of those towns experiencing an expanding economy and a growing demand for labour – hence,
potentially a place with good opportunities for "selling" unemployed job
applicants for local enterprises. However, even in such a town, Majoriaq

employees occasionally have a hard time goading local companies into on-boarding local workers.

> I've made the unemployed [clients] apply for job in [company name]. Thirty-three applications have been sent off—not one has been hired. The same company appreciates their applications, but they don't hire them. A month ago, this same company applied for permission to import three workers from [country of origin of this company]. A carpenter, a mechanic and a handy man, and we have those people here, but they prefer their own. So I email them that we got local people, why can't you use them? I stop processing their application and ask for stronger arguments.

This employee has chosen what we term the offensive strategy towards the companies. Respondents from other towns tell similar stories that illustrate a particular characteristic in relation to the Greenlandic labour market. Outside Nuuk, fishing, the service sector, and mining tend to dominate the labour market. Some companies (e.g., in mining) have foreign owners and managers, and many of those companies regard local workers as unstable and badly educated. Therefore, such companies tend to prefer importing foreign workers. Still, such import requires an application for Majoriaq, and the company needs to argue for its necessity. The application requires the Majoriaq's approval before the company can go ahead. This approval process brings the employees of the Majoriaq into a situation that they may try to exploit offensively to coax the companies into hiring (more) local workers. According to different Majoriaq employees, the companies usually obtain the requested permission to import foreign labour. Still, by emailing the companies a list of available local unemployed workers, the employees seek raising the awareness of the employers of local options instead of foreign workers. One employee said, "When vacant positions show up in the companies, we email the company a list of qualified unemployed citizens. When the company, all the same, applies for permission to import foreign labour, at first we say no".

Managers and employees also tried other ways to adopt an offensive approach. In a medium-sized town, the Majoriaq manager sought to change the usual layoff pattern of the largest company in the town, a fish-processing factory. The factory usually lays off workers temporarily when there are no fish. However, these dismissals are a major administrative burden to the employees in the Majoriaq and have a negative impact on municipal finances. Therefore, the manager tried to convince the company that instead of dismissing the workers, they should provide these workers with in-service training. The manager said, "We negotiate with [the fish-processing factory] on not dismissing the workers during winter and instead providing these workers with some training during low season".

In other towns, Majoriaq managers and employees approached the local enterprises in a more defensive manner. They argued that, typically, benefit recipients from Match Group 1, are able to find job openings themselves; as to the employers, they also know how to use local social networks to find qualified workers (cf. Granovetter, 1995). Such an approach among the caseworkers had an element of coping to it (cf. Lipsky, 1980), because we mainly found this approach in small towns where the Majoriaq has few employees who need to prioritise some tasks over others. In these towns, there are no traditions for seeking to negotiate placements with the companies. Typically, the caseworkers will just send an email to different companies in the case of seeking to assist an unemployed citizen to find a job instead of visiting or negotiating with the companies.

4.9.2 Collaboration concerning development of the work capacity of vulnerable workers

Another important field of collaboration between the Majoriaq and the local workplaces centres on work capacity testing and disablement rehabilitation jobs. Depending on whether the focus was on public (municipal) or private workplaces, managers and employees' views on this topic differed. Testing the work capacity of an unemployed citizen is costless to a company, but a company must pay a gradually increasing share of the salary of a worker on a disablement rehabilitation contract. According to our interviewees, compared to private companies in some towns, municipal workplaces are more willing to take part in efforts to enhance the employability of vulnerable workers. Two managers from two small Majoriaq said, respectively:

> It is difficult to make private companies take in disablement rehabilitation workers because they cost a bit. Therefore, these workers end up in public workplaces where the communal cash box finances the salary. Private companies are happy to stage work capacity tests because, to a company, such tests are costless.
>
> Private companies are unwilling to take in unemployed citizens who need to undergo a job clarification process. Those unemployed citizens who need a work capacity test end up in a municipal workplace. Also for disablement rehabilitation jobs, we cannot find enough positions. When we succeed, they sack the citizen in the end, and then we need to start over again.

However, in other towns, typically those with an expanding economy, interviewees expressed the opposite view, that is, that collaboration worked better in private workplaces than public ones. A manager and an employee from two large Majoriaq said, respectively:

> We collaborate well with private companies but badly with municipal workplaces, because municipal workplaces require a clean criminal record, and they have no job openings, not even for a flexi-job.

Workplaces rarely keep workers with a disablement rehabilitation contract. The job ends when the citizens start costing money. Municipal workplaces tend to be the worst ones. They are nice and socially encompassing, but they never hire.

In the survey of Majoriaq employees, we asked how they would rate collaboration with private companies pertaining to: (a) jobs for unemployed persons from Match Group 1, (b) positions for persons who need a disablement rehabilitation job, (c) jobs for persons who have been through disablement rehabilitation. Mainly, the employees expressed the view that collaboration worked well in relation to persons from Match Group 1. They were slightly less positive in terms of their views on locating positions for persons who needed a disablement rehabilitation job. The least favourable ratings concerned finding jobs for persons who had been through disablement rehabilitation. Still, on a 5-point scale, 50% rated collaboration on this issue as working "well" or "very well", 44% found collaboration to be working "fair", while 6% rated it as "bad" or "very bad".

Summing up, unsurprisingly, collaboration with public and private companies works better when job applicants are job ready and possess the skills that a private or public workplace requires – compared to the opposite situation. This reflects an employers' logic that they, for obvious reasons, prefer the applicant that they find most suitable for the position. Still, an adverse consequence of this logic in Greenland's case is that many private companies prefer importing workers from their own country of origin over hiring workers with a Greenlandic background. To both unemployed local workers and employees in the Majoriaq, this is a competitive challenge that may leave local workers stranded in unemployment. An even bigger challenge lies in finding positions for unemployed workers from Match Group 2, who need either a disablement rehabilitation job or, building on a foregoing rehabilitation process, an ordinary job or a flexi-job (Antropologerne, 2016). Still, even concerning the latter thorny task, the Majoriaq employees in our survey rated collaboration with the local enterprises to be working relatively well.

4.10 Conclusions

A challenging matrix of demography and geography – few people dispersed on the lengthy fringes of an enormous ice cap – characterizes the labour market in Greenland. Nuuk, along with Sisimiut and Ilulissat, are the centres of economic growth, while many smaller towns struggle with stagnating local economies and high unemployment rates. A labour market institution like the Majoriaq with the goal of facilitating good matches between unemployed workers and labour-seeking employers operates under difficult conditions when trying to realize this goal. Aggravating the challenge is that, compared to the other Nordic countries, the working-age population in Greenland has a relatively low level of education, while more and more jobs require increasingly advanced skills. Some of the Majoriaq employees interviewed had experienced many companies, mostly foreign but also some Greenlandic, preferring foreign workers to workers from Greenland, as they perceive these workers to be better educated, more efficient and/or more stable. This is a hitch – both to unemployed job-ready workers whose only problem is that they lack a job and to the Majoriaq whose employees need to invent strategies to spur companies to hire local workers.

During recent years, Greenland, like many other Western countries, has started focusing on how to put more of the "hard-to-place" unemployed benefit recipients to work. Those are the unemployed benefit recipients in Match Groups 2 and 3, and some of those recipients whose early retirement pension entitlements will be reassessed and potentially revoked. Most Western countries have limited success with putting the most vulnerable populations to work (Andersen et al., 2017), and the structural conditions characterizing the Greenlandic labour market make realizing this goal even more difficult there. Recent labour market legislation has introduced procedures and programs that should further the employability of benefit recipients in Match Group 2 and of some of the functioning benefit recipients in Match Group 3. Although psychologically demanding, Majoriaq managers and employees tend to regard the job clarification process, including the CM course, as a useful tool in paving the way towards the labour market for these benefit recipients. Still, the barriers to overcome are non-negligible. One barrier is the depth of psychological problems and trauma characterizing some of these benefit recipients. A second barrier is the relative lack of competencies to handle such problems among the many clerically educated employees in the Majoriaq labour market sections. A third barrier is the occasionally difficult collaboration between the Majoriaq and the municipal social services departments. Finally, companies are even more reluctant to on-board these workers, compared to job-ready workers, although many companies are willing to participate in work capacity tests and provide disablement rehabilitation jobs.

As should be evident, the challenges facing Greenlandic politicians and civil servants in the employment field are still serious. The problems need to be addressed long before citizens turn up in the unemployment queue because the foundations for capabilities leading to success in the labour market are laid during early childhood – a period during which neglect may leave negative traces that are difficult to repair later on (Cunha et al., 2006; Knudsen et al., 2006). Public health surveys show that violence against children and alcohol abuse among parents in Greenland are decreasing (Larsen et al., 2019). Still, many children in Greenland – especially in some remote towns and townships – grow up in homes characterized by violence, abuse, and neglect. Decision makers need to address such problems through adequate family policies targeting troubled families and children nationwide.

A three-week CM course in the Majoriaq later in life – even with follow-up meetings during the ensuing year – is unlikely to repair the potential damage fully. The same goes for educational policies that need to improve on a school system where too many young people fail to achieve the public school diploma and hence lack the most basic qualifications for either an unskilled blue-collar job or the pursuit of higher education.

Focusing specifically on the employment system, our research shows that employees in the Majoriaq need more support because their skills are insufficient for the challenges they encounter. In a short-term perspective, more supervision and a travel team of psychologists and social workers may provide support for some of the most hard-pressed clerical workers in the Majoriaq. In a long-term perspective, more psychologists and formally educated social workers are required in either the Majoriaq themselves or in the municipalities to assist Majoriaq employees. Citizens who have experienced sexual abuse or violence during their childhoods and/or who struggle with alcohol or substance abuse need adequate counselling, treatment, and therapy. There is also a need to convince more companies to consider employing more workers with a Greenlandic background and especially workers who have had a disablement rehabilitation job and now need an ordinary job or a flexi-job. In some towns, local politicians, civil servants from the Majoriaq, and representatives from the local business community meet to discuss and find solutions to the problems characterizing the local labour market. Such local consultations may be a way forward for increasing the inclusiveness of the Greenlandic labour market—in all its local and more or less insular incarnations.

Notes

1. For more details on the unemployment benefits scheme, see www.sullissivik.gl/Emner/Arbejde/Arbejdsmarked/Arbejdsmarkedsydelser?sc_lang=da-DK.
2. For a more details on the Greenlandic public benefits scheme, see www.sullissivik.gl/Emner/Sociale_forhold/Offentlig-hjaelp/Offentlig_hjaelp_Generelt_om?sc_lang=da-DK.
3. For more details on the early retirement pension scheme, see https://www.sullissivik.gl/Emner/Sociale_forhold/Foertidspension/Foertidspension_nyt-design?sc_lang=da-DK.
4. In Denmark in 2019, almost 89,000 persons held a flexi-job equivalent to 2.8% of the workforce (own calculations, based on data from www.jobindsats.dk).
5. Please remark that for 2017, Statistics Greenland (2020, p. 15) enumerates 5,788 recipients of public benefits, 2,912 recipients of unemployment and disease benefits, and 2,280 recipients of early retirement pension. The differences reflect different methods for calculating the number of recipients. While the method used in this chapter calculates a monthly average of recipients relative to the different types of benefits, Statistics Greenland enumerates the number of recipients who have received a certain type of benefit during the year.

References

Agresti, A. (2018). *Statistical methods for the social sciences* (5th ed.). Pearson Education.

Andersen, N. A., Caswell, D., & Larsen, F. (2017). A new approach to helping the hard-to-place unemployed: The promise of developing new knowledge in an interactive and collaborative process. *European Journal of Social Security, 19*(4), 335–352.

Antropologerne. (2016). *Udfordringer og muligheder for ledige: En undersøgelse af lediges behov samt årsager til ikke at møde op til anvist arbejde* [Challenges and Options for the Unemployed: A Investigation on the Reasons among Unemployed Citizens for not turning up for Assigned Work]. Government of Greenland.

Askim, J., Fimreite, A. L., Moseley, A., & Pedersen, L. H. (2011). One-stop shops for social welfare: The adaptation of an organizational form in three countries. *Public Administration, 89*(4), 1451–1468.

Barr, B., McHale, P., & Whitehead, M. (2019). Reducing inequalities in employment of people with disabilities. In U. Bültmann & J. Siegrist (Eds.), *Handbook of disability, work and health* (pp. 1–19). Springer.

Boolsen, M. W. (2012). *PIAREERSARFIIT i krydsfeltet mellem arbejdsmarkedspolitik, uddannelsespolitik og vejledning* [PIAREERSARFIIT at the crossroads of labor market policy, education policy and guidance]. Institut for Statskundskab.

Bredgaard, T. (2011). When the government governs: Closing compliance gaps in Danish employment policies. *International Journal of Public Administration, 34*(12), 764–774.

Broberg, U. (2019). *Rapport over Vejlednings—Og Motivationsforløb* [Report on Guidance and Motivation Courses]. Departementet for Råstoffer og Arbejdsmarked.

Cunha, F., Heckman, J. J., Lochner, L., & Masterov, D. V. (2006). Chapter 12: Interpreting the evidence on life cycle skill formation. In E. Hanushek & F. B. T. (Eds.), *Handbook of the economics of Welch education* (1st ed., pp. 697–812). Elsevier.

Granovetter, M. S. (1995). *Getting a job: A study of contacts and careers* (2nd ed.). University of Chicago Press.

Holt, H., Thuesen, F., & Casier, F. (2019). *Evaluering af Majoriaq-centrenes arbejdsmarkedsindsats* [Evaluation of the labor market efforts of the Majoriaq centers]. VIVE—Det Nationale Forsknings- og Analysecenter for Velfærd.

Karagiannaki, E. (2007). Exploring the effects of integrated benefit systems and active labour market policies: Evidence from Jobcentre Plus in the UK. *Journal of Social Policy, 36*(2), 177–195.

Kjeldsen, L. (2020). Jobcentrenes koordinering med interne og eksterne aktører [Jobcentre coordination with internal and external actors. In T. Bredgaard, F. Amby, H. Holt, & F. Thuesen (Eds.), *Handicap og Beskæftigelse: Fra Barrierer Til Broer* (pp. 309–332). Djøf Forlag.

Knudsen, E. I., Heckman, J. J., Cameron, J. L., & Shonkoff, J. P. (2006). Economic, neurobiological, and behavioral perspectives on building America's future workforce. *Pnas, 103*(27), 10155–10162.

Knuth, M., & Larsen, F. (2010). Increasing roles for municipalities in delivering public employment services: The cases of Germany and Denmark. *European Journal of Social Security, 12*(3), 174–199.

Kvale, S. (1996). *Interviews: An introduction to qualitative research interviewing.* SAGE Publications.

Larsen, C. V. L., Hansen, C. B., Ingemann, C., Jørgensen, M. E., Olesen, I., Sørensen, I. K., Koch, A., Backer, V., & Bjerregaard, P. (2019). *Befolkningsundersøgelsen i Grønland 2018: Levevilkår, livsstil og helbred* [Population survey in Greenland 2018: Living conditions, lifestyle and health]. SIFs Grønlandsskrifter, nr. 30. Statens Institut for Folkesundhed.

Larsen, E. L., Jensen, J. M., & Pedersen, K. M. H. (2020). Cross-sectorial collaboration in return to work interventions: Perspectives from patients, mental health care professionals and case managers in the social insurance sector. *Disability and Rehabilitation*, 1–8. DOI:10.1080/09638288.2020.1830310.

Lipsky, M. (1980). *Street level bureaucracy: Dilemmas of the individual in public services.* Russel Sage Foundation.

Pedersen, N. J. M., Petersen, J. S., & Lindeberg, N. H. (2018). *Analyse af offentlig hjælp* [Analysis of public assistance]. VIVE—Det Nationale Forsknings- og Analysecenter for Velfærd.

Portes, A. (1998). Social capital: Its origins and applications in modern sociology. *Annual Review of Sociology, 24*(1), 1–24.

Smylie, M. A., & Evans, A. E. (2006). Social capital and the problem of implementation. In M. I. Honig (Ed.), *New directions in education policy implementation* (pp. 187–207). State University of New York Press.

Statistics Greenland. (2020). *Greenland in figures 2020.* Statistics Greenland.

Winter, S. (2014). Implementation. In B. G. Peters & J. Pierre (Eds.), *The SAGE handbook of public administration* (2nd ed., pp. 227–236). SAGE.

5　Mapping the Greenlandic labour supply

Evidence from Greenland's first labour force survey

Cecilie Krogh and Laust Høgedahl

Abstract

A good match of supply and demand of work in any labour market is of importance to the overall economy and the quality of work. Therefore, knowledge concerning these two components is essential to policymakers and other stakeholders. This chapter aims to investigate the supply side of the Greenlandic labour force. The investigation includes a new quantitative measurement of employment and unemployment rates. Further, the chapter examines time spent at work and the satisfaction of work and length of time employed at one workplace. Finally, the chapter includes an examination of the self-employed in Greenland. The data stems from Greenland's first national Labour Force Survey, in which 1,000 Greenlanders participated. The methodology includes a standardised questionnaire allowing an international comparison of the results. Our results show a low unemployment rate, a high number of weekly working hours, and high job satisfaction. At the same time, the results show a surprisingly low seniority in Greenland and a fairly large share of the working population not part of the workforce, thus labelled as inactive. Regarding the self-employed, we find a homogeneous group consisting almost only of middle-aged males fishing from small open boats. We conclude our chapter by listing a number of key issues that need the attention of policymakers and other stakeholders in order to raise the quantity and quality of the labour supply.

5.1 Introduction – new light on Greenland's labour supply

Any labour market basically consists of a supply side and a demand side (Abbott & Ashenfelter, 1976). The demand side stems from employers (public and private), while the supply side includes the current labour available. For the past 20 years, most Western countries have developed Labour Force Survey data to complement and supplement existing register or administrative data on the labour supply. These data gives policymakers and other stakeholders useful insights in terms of, for example, unemployment levels and motivation for work. In 2019, the first beta version of the Labour Force Survey conducted in Greenland was aimed at persons between 16 and

64 years old and provided unique quantitative data representative of the entire Greenlandic workforce.

For the past years, the match between supply and demand has been heavily debated in Greenland. Employers have voiced concerns in terms of labour shortages creating bottleneck situations in various industries. The labour shortages are seen even though a rather large share of the Greenlandic population between 18 and 65 years of age, for various reasons, is not actively participating in the labour market. The labour shortage will add costs to planned construction projects and hamper the country's overall economic growth in the future (Government of Greenland, 2020). The question is what can be done? Some are calling for more easy access to foreign workers, others point to lowering social benefits in order to create motivational effects, and still others are focusing on a more active labour market – and educational policies (see Chapter 4).

In this chapter, we set out to analyse the modern Greenlandic labour supply by looking at key features such as unemployment levels, weekly working hours, and seniority/mobility. Because the data is comparable to other Labour Force Surveys, we are able to compare the Greenlandic results with other countries and ask the following question: What differences and similarities are there between the Greenlandic labour market supply and the labour market supply of other countries? This gives us a unique opportunity to investigate and test how the Greenlandic labour market supply looks in an international light. These findings are very much absent in the existing literature, which mainly consists of qualitative studies, as we will outline below.

We begin the chapter by presenting a review of the existing literature dealing with the Greenlandic labour supply. By addressing this interesting and solid body of work, we are able to identify key features associated with the Greenlandic labour market. In this section, we also draw on international findings relevant when studying the Greenlandic labour supply mainly in comparison to other countries heavily dependent on fishing and its downstream industries. We then turn to the methods and data used in our analysis followed by the results from the first Greenlandic Labour Force Survey. We conclude by summarising differences and similarities when comparing the Greenlandic data to that of other countries. We find that the Greenlandic labour supply in many ways is rather unique, but we also find some interesting similarities with other countries. In the final section, we point to some areas of special interest that call for more research and attention from policymakers.

5.2 Creating a supply and demand for labour – the transformation from a self-sufficient to a production-based economy

Traditionally, most Greenlandic families have been self-sufficient (self-employed) in fishing, hunting, and trading. After World War II, the Greenlandic economy slowly transformed into a production-based economy with a supply and demand for labour as a consequence (Thorleifsen, 2005; Chapter 3). However, in the 1920s, fishing had already surpassed whaling and sealing as the main area of employment in Greenland, but from 1950 onwards, even more Greenlanders found employment in the "new economy", based mainly on the downstream industries connected to commercial fishing. By 1955, the share of employees in Greenland was 40% (60% self-employed) compared to 15.7% in 1930 (84.3% self-employed). The first employees were predominantly employed at KGH – Royal Greenland Trading [Den Kongelige Grønlandske Handel], a Danish state-owned company trading provisions and goods between settlements around Greenland (Marquardt, 2005).

Since the 1950s, the Greenlandic labour supply has grown exponentially. A rather large share of Greenlanders still make a living as self-employed – mainly men fishing solo from small open boats (Høgedahl & Krogh, 2020). The rather late industrialisation of the Greenlandic economy, compared to other Western European countries, also came with new challenges in terms of matching the supply and demand of labour. Education and labour market policies were needed to supplement the industrial policy (Thorleifsen, 2005). The need for education policies is still very present today.

During the transition period from a colony to an equal territorial part of the Kingdom of Denmark, more specialists and workers with managerial rights from Denmark took foothold in the Greenlandic labour market, suppressing native Greenlanders (Bianco, 2019). This process caused an increasing dualisation of the labour supply where prominent positions for the most part were occupied by Danes and to a lesser extent native Greenlanders. Today, the modern labour supply is still highly segmented in Greenland. This trend has been even stronger because many employers, especially within the dominant fishing industry, have argued in favour of easier access to foreign workers. Chinese seasonal workers especially have since taken unskilled work in Greenland during peak seasons.

Research and data concerning the Greenlandic labour supply have been in demand for many years (Government of Greenland, 2020). In spite of having a rather small domestic workforce of around 33,000 persons, the Greenlandic labour supply is addressed in some studies we found when surveying the literature. The research tends to fall into three categories:

1 Motivation and mobility
2 Barriers in terms of labour market participation
3 The composition of the Greenlandic labour supply

Overall, most research has a point of departure in qualitative methods often based on anthropological or ethnographic approaches (Antropologerne, 2016). In terms of motivation and mobility, we find a number of studies. Petersen (2005) shows a great deviation between employers and employees when it comes to motivational factors for taking up paid work in Greenland.

Many employers find the work both interesting and mentally rewarding, whereas employees are motivated mainly by the social aspect and comradeship with co-workers. The disparity in values is formed by multiple factors including culture differences between native Danish managers and native Greenlandic workers (Petersen, 2005). Research also indicates different motives among men and women creating barriers for labour market participation, especially among men (Poppel, 2005; Prattis & Chartrand, 1990). A long tradition as whalers and sealers has created a strong identity for many male Greenlanders as breadwinners providing for their families by hunting and fishing. Many men consider factory work a female occupation. Because the downstream industry from commercial fishing is dominant in the Greenlandic labour market, the lack of male participation is problematic and challenging.

5.3 The Greenlandic labour force survey

The Labour Force Survey is conducted in most countries around the world following the standards set by the International Labour Organization. In the European context, the Labour Force Survey is currently conducted in all member states of the European Union, four candidate countries, and three countries of the European Free Trade Association in accordance with Council Regulation No. 577/98 of 9 March 1998. The European Union Labour Force Survey is a large household sample survey (cross-sectional study) providing for most countries quarterly results on the labour participation of people considered of working age (16 to 65), including persons outside the labour force living in private households.

The Labour Force Survey data describes the labour market status of the population. The Labour Force Survey gives insight into how many people are employed, unemployed, or outside the labour force (economically inactive). These numbers can be hard to provide solely by looking at register data. The Labour Force Survey also provides information comparing how many people are working part-time, how many hours people usually work, and what their connections are to the labour market. The Labour Force Survey can thus provide more factual information in addition to attitudes towards labour market issues.

A preliminary version of the Labour Force Survey was conducted for the first time in Greenland in March 2019 by Aalborg University, Denmark. The survey was collected by a Greenlandic consulting agency from the 8th of April to the 22nd of May 2019, and the sample consisted of 1,000 respondents. Ahead of data collection, the survey was pilot tested on 100 respondents, and the test showed the need for only a few changes. The responses from the pilot test were collected separately and do not figure into the final data. The respondents were randomly sampled from the customer list of Tele Greenland. After the data collection, the realised sample was tested against available register data from Greenland Statistics in terms of gender, age, and geography. The data proved to be very representative, and no additional adjustments through weights were needed.

The data collection followed the computer-assisted telephone interviewing method (Groves et al., 2009, p. 151). The respondents were interviewed by phone, with the interviewer reading the questions out loud and typing in the answers while interviewing. The survey was available in Greenlandic and Danish. A minority of the Greenlandic population (roughly 15%) speak only Danish, 70% speak only Greenlandic, and 15% are bilingual, speaking both Greenlandic and Danish (Saammaateqatigiinnissamut Isumalioqatigiissitaq, 2017). Of the respondents, 92.6% answered the questionnaire in Greenlandic. For more on the data collection, please see Høgedahl and Krogh (2020).

5.4 The labour force survey measuring of employment, unemployment, and the inactive population

The Labour Force Survey uses a specific model for measuring the labour market participation of the population. This model makes the results internationally comparable. All populations receive the same questions, and when the questions are processed equally, the results should be comparable across nations. In this section, we present the definition of employed, unemployed, and the inactive population in terms of the Labour Force Survey.

In the Labour Force Survey, most questions ask the respondent for their labour market participation in a specific week called the reference week. The reference week in this study is 25–31 March 2019. By asking all respondents about labour market participation in the same week, we ensure that their circumstances are the same. For example, the structural circumstances can easily change in a short time. As seen with COVID-19, the situation of the labour market can change very fast. When using a reference week, all respondents answer the questions within a comparable time interval. By using a reference week, we ensure that all responses are comparable.

Eurostat (2020c) defines the employed, unemployed, and inactive as follows,

> **Employed persons** are persons aged 15 years and over who, during the reference week, performed work even for just one hour a week, for pay, profit or family gain or who were not at work but had a job or business from which they were temporarily absent because of something like, illness, holiday, industrial dispute or education and training.
>
> **Unemployed persons** are persons aged 15–74 who were without work during the reference week, but who are currently available for work and were either actively seeking work in the past four weeks or had already found a job to start within the next three months.
>
> The economically **active population** comprises employed and unemployed persons.

Inactive persons are those classified neither as employed nor as unemployed.

To be defined as employed, the respondent needs to be either working or absent from work temporarily during the reference week. Temporary absences include those for short-term illnesses or other reasons forcing the respondents to stay at home temporarily.

This study includes persons of ages other than those in Eurostat's definitions. The ages of the employed and unemployed differ between countries due to different national retirement ages, and a retired individual should not be included in the categorisation of labour market participants. The study of Greenland follows the same age categories as Denmark. The age is therefore not 15–74 years but 16–64 years. However, the survey does not include any respondents under the age of 17. Regarding retirement, respondents under the age of 64 can still be retired. This is taken into account in the questionnaire.

As explained in the introduction to this chapter, there are various estimates of unemployment rates. The estimate becomes important in the question of labour supply because it provides knowledge about the potential labour supply. In this section, we will define the labour force in Greenland by using the Labour Force Survey methodology and making an international comparison. We compare the results from Greenland with the average results of the European Union, Denmark, and Iceland. As Greenland is part of the Kingdom of Denmark, we found it interesting to compare the results. The comparison with Iceland was chosen because of the similarities between the countries. Both countries are islands in the North Atlantic and consist of relatively small workforces with a dominant fishing industry that is important to the national economy. Furthermore, Greenland and Iceland share a colonial history.

As Table 5.1 shows, the majority of the Greenlandic people were employed during the reference week in March 2019. Of the respondents, 78.3% report working in the reference week and 21.7% reported not having a job in the reference week.

We further divide the latter group of respondents into two subgroups: (a) respondents that did not have a job in the reference week but did apply for a job within the last four weeks and (b) respondents that did not have a job in the reference week and did not apply for a job within the last four weeks.

The first subgroup accounts for 5.4% of all respondents. Subgroup 1 is divided further into two groups: those who can start working within 14 days and those who cannot. Figure 5.1 shows only the groups of respondents who are able to start working within 14 days.

Due to this method, we define only 2.6% as unemployed in the same reference week. Only a few individuals are therefore categorised as unemployed in Greenland.

Table 5.1 also reveals that some of the respondents are neither employed nor unemployed in the reference week, the so-called *inactive population*.

Table 5.1 Employment and unemployment in Greenland in 2019

Did not have a job in the reference week	*Did have a job in the reference week*
21.7%	*78.3%*
Subgroup 1. Did not have a job in the reference week but did apply for a job within the last four weeks	Subgroup 2. Did not have a job in the reference week and did not apply for a job within the last four weeks.
5.4%	16.3%
Did not have a job in the reference week but did apply for a job within the last four weeks. The individual states that they can start working within 14 days.	
(LFS unemployment rate)	
2.6%	

Source: The Greenlandic Labour Force Survey.

The inactive population are individuals who did not have a job in the reference week and either did not apply for a job within the last four weeks (Subgroups 2) or are not able to start working within 14 days (approximately half of the individuals of Subgroup 1). The inactive population counts for 19.2% of the respondents. Based on these numbers, the measurement of employment and unemployment rates in Greenland is therefore as shown in Figure 5.1.

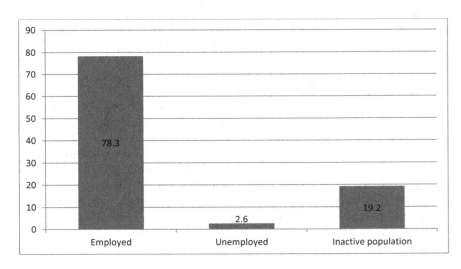

Figure 5.1 Employment and unemployment rates in Greenland 2019 (in %).
Source: The Greenlandic Labour Force Survey.

Converted to absolute numbers, the results show that 991 individuals aged 17–64 are unemployed in Greenland, 7,320 individuals are inactive, and 29,852 individuals are employed. We might expect rather large differences across the country due to the highly segmented labour market, as described above. In any case, there can be several reasons for being periodically or permanently inactive in the labour market. Reasons can relate to the individual and be caused by an interplay of different variables like health problems, lack of education, age, or other individual factors (see also Chapter 7). Individual factors are sometimes changeable, for example, lack of education or other training and skills. On the other hand, structural factors can cause rising unemployment rates. Economic recessions, such as the ones from 2008 to 2015 or during the recent COVID-19 crisis, reduce the demand for labour and challenge both national and local economies.

The reduced job demand can cause a rise in unemployment rates. A rise in unemployment rates not only increases public expenses but it also can have a significant social and economic impact on individuals. In modern society, work is not just a source of income but also a space for social network and identity (Kenny et al., 2011; Leidner, 2006). The loss of a job can also mean the loss of identity (Jahoda et al., 1971; Price et al., 1998).

Furthermore, economic recessions affect societal groups differently. Studies from Denmark show that already vulnerable individuals like unskilled workers and immigrants are more likely to lose their jobs during the COVID-19 crisis than those who were higher educated (Danish Agency for Labour Market and Recruitment, 2020).

In our study, individual factors are of concern. Regarding the 19.2% inactive population, of those who were not able to start working within 14 days, one-fifth state that health problems cause their situation. Another fifth are unemployed due to forthcoming graduations. For those who are characterised as inactive because they did not apply for a job within the last four weeks, 34.1% state that it is because of forthcoming graduations. Due to disability or illness, 9.6% did not apply for a job within four weeks. This might seem like a relatively low share; however, 19.3% did not apply for a job because they already receive or have applied for early retirement benefits. To receive early retirement benefits, one must document a decreased work capacity.

There is a great disparity between the explanations for being inactive in the labour market. We might expect the group with forthcoming graduations to be included in the labour force and gain employment in a foreseeable time. However, at the moment, they are not expected to be in the labour market. It is a more complex situation with individuals who are inactive due to health issues. They are not expected to enter the labour market unless their health situations improve or workplace environments become more inclusive. Labour market inclusiveness could be part-time work, employment with wage subsidies, sheltered employment, or other initiatives that aim to integrate persons with different health issues in the labour market.

If the health problems relate to disability or other long-term health issues, health improvements are not necessarily achievable. In such situations, the question is whether workplace environments can be adapted to the challenges experienced by these individuals.

Even though health issues do not necessarily correlate with social, economic, or educational issues, previous research documents that individuals from low socio-economic positions are far more likely to experience health issues than individuals from high socio-economic positions (Eisenberg-Guyot & Prins, 2019; Fein, 1995; Mackenbach, 2006; OECD, 2019). Studies name this trend *health inequality,* a widespread phenomenon. The situation of the group with health issues may therefore be more complex and not related only to health. In this study, the combinations of unemployment and health issues are at least at stake. Based on previous research, there could be other issues affecting the situation.

Only 7.2% of the respondents are unable to take on a job within 14 days because of obligations related to their families.

Little more than half of the individuals who cannot take on a job within 14 days state reasons other than those above. The same goes for 15.6% of the individuals who did not apply for a job within four weeks before the reference week. The results do not provide a clear answer on the other reasons. However, they show a complexity in the causes of inactivity, and further research could look into this aspect. Hence, there is a potential to increase the labour force supply from the 19.2% inactive population. Almost one fifth of the full working-age population is inactive in the labour market and cannot be labelled as unemployed. To improve the understanding of the potential, knowledge about the individuals who are not employed is necessary. Our results show that for some, health issues or forthcoming graduations cause the situation; however, there are other unexplained causes at stake as well (see also Chapter 4). One might speculate that the rather large share is also connected to more structural and cultural explanations. If Greenlanders are able to provide for themselves by hunting and fishing, they might not be able to take up work within 14 days.

Compared globally, the Greenlandic unemployment rate of 2.6% is low (cf. Table 5.2) In 2019, the unemployment rates of countries completing the

Table 5.2 Employment, unemployment, and inactive population in Greenland, EU countries, Denmark and Iceland, 2019 (in %)

Country	LFS employment rates (%)	LFS unemployment rates (%)	LFS inactive population (%)
EU (28 countries)	69.2	6.3	25.8
Denmark	75	5	20.6
Iceland	84.1	3.5	14.3
Greenland	78.3	2.6	19.3

Source: Eurostat 2020a.

Labour Force Survey ranged between 17.3% in Greece and 2% in Czechia. Compared with the average of the EU, Denmark and Iceland, the Greenlandic unemployment rate is also low.

The share of inactive individuals in Greenland is also lower than the average of the European Union countries and Denmark. On the other hand, the share in Iceland is lower than the share in Greenland.

Altogether, the first part of the results shows a high employment rate and low unemployment rate in Greenland in 2019. Even so, the results also point to some potentials for increasing the labour supply.

5.5 Labour supply

After defining the employment and unemployment rates in the previous section, we now turn to the employed population. In this section, we explore the amount of work employed individuals in Greenland perform. If a political goal is to increase the labour supply in Greenland, where the possibilities are must be investigated. As mentioned in the previous section, the unemployment rate in Greenland is rather low, but there might be potential in the inactive population, that is, people not seeking work or not able to take up work within 14 days. Another way to increase the labour supply is by increasing the labour supply of individuals already employed.

Regarding the amount of work among people already employed, the results show a high number of weekly working hours in Greenland. A normal workweek in Greenland consists of 40 hours. In comparison, Denmark defines a full-time workweek as 37 hours.

As shown in Figure 5.2, a 40-hour workweek is most common in Greenland. The figure also shows the difference between workers with and without an agreement on working hours. Blue-collar workers with hourly wages will, in most cases, have a contract stating a 40-hour workweek. White-collar workers are paid mainly by the month and do not, in most cases, have a 40-hour workweek. Not surprisingly, workers (mainly blue-collar workers) with an agreement are more likely to work 40 hours a week than workers without an agreement. The majority of workers (72.9%) have an agreement with their employer about weekly working hours.

Figure 5.2 shows the respondents' normal, or agreed upon, working hours in a random week. When asked about their actual working hours in the reference week, the respondents claimed they work 38.1 hours on average. In comparison, the average number of working hours in Denmark in the first quarter of 2019 was 33.8 hours (Eurostat, 2020b). That is a difference of 4.3 hours per week – almost one hour a day during working days. Compared with the average of the 28 European countries, the number of working hours in Greenland is also relatively high. In the first quarter of 2019, the average number of working hours in the 28 European countries was 36.3 hours per week, 1.8 hours less than in Greenland (Ibid.). In Iceland, the average number of working hours was 37.9 per week, close to the Greenlandic result (Ibid.).

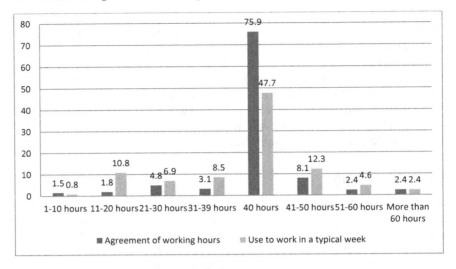

Figure 5.2 Working hours in a typical week split on individuals with and without an agreement of working hours, 2019 (in %).

Source: The Greenlandic Labour Force Survey.

One explanation for the high number of working hours in Greenland is that only 10% of the population works part-time. In Europe, 19.2% of the population worked part-time in 2018 (European Commission, 2019). In Denmark, the share of part-time workers in 2019 was 24% (DST, 2019). The high share of people working part-time decreases the average number of working hours, which is not the case in Greenland. The relatively small share of persons working part-time could be a reason for the high number of working hours.

The high number of working hours and small number of part-time workers in Greenland leaves a small amount (if any at all) of room for increasing the labour supply among already employed individuals. It is difficult to imagine that workers in Greenland can increase their average number of working hours significantly.

In spite of spending much time at work, Greenlandic workers on average have a very short seniority. Nearly 60% of the workers have worked at their present workplace since 2016. That is only three years at the same workplace.

Figure 5.3 shows 9% of workers have worked at their current workplace for more than 20 years. On average, Greenlandic workers have worked at the same workplace for 6.6 years. That is a very short time compared to other countries. Denmark is known for its high job mobility (Madsen, 2004); in 2018, an average Danish worker had worked at the same workplace for 7.5 years (OECD, 2018). That is almost a year longer than Greenlandic

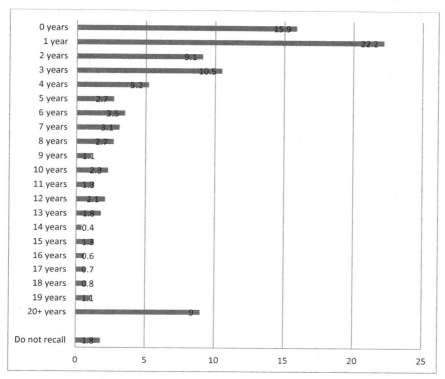

Figure 5.3 Years at present workplace, 2019 (in %).
Source: The Greenlandic Labour Force Survey.

workers. In Iceland, an average worker has worked at the same workplace 1.6 years longer than the average Greenlandic worker. The average Icelandic worker in 2017 had worked at the same workplace for 8.2 years (Ibid.). In 2018, the average seniority in the OECD countries was 10.2 years (Ibid.).

High job mobility is often considered a positive trait of any labour market, implying a strong degree of flexibility. Job shifts can be understood as part of a rising career or at least as an individual opportunity (Moscarini & Thomsson, 2008; Rosenfeld, 1992). Building up a career is not only about job shifts but the shifts can imply better job opportunities. However, in some cases, high job mobility can be problematic, especially for employers because they can experience transition costs linked to recruiting and training new employees. Moreover, high job mobility might translate to a loss of human capital. However, when is job mobility too high? Without putting a number on the phrase "too high", we find that the Greenlandic workers have high job mobility compared to other countries and that high job mobility can cause challenges for employers, as also mentioned in other recent studies.

Explanations for the high job mobility can relate to both structural circumstances, employers dismissing employees, and employees leaving the workplace.

The fishing industry in Greenland is large, and many workers and companies are dependent on the flow of fish stocks (Høgedahl & Krogh, 2020). This is found to be a reason for the dismissal of workers, but they are rehired when the work returns (Chapter 4). If the respondents are subject to this "firing and hiring" strategy, it might explain part of the high job mobility. However, this is only an explanation as long as the respondents are hired at a workplace other than the one from which they were dismissed. If the respondents are rehired at the same workplace, they are most likely expected to possess work experience from before their dismissal.

The high job mobility can also indicate difficulty retaining employees at workplaces. The question is why these difficulties occur. Research indicates a relation between high job mobility and individual dissatisfaction connected to the job situation (Kristensen & Westergård-Nielsen, 2004). Individuals who are less satisfied with their jobs are – not surprisingly – more likely to "vote with their feet", that is, to quit. Using one's voice to speak out and exiting a job are the most applied strategies when workers experience dissatisfaction with their working lives. One might therefore expect that Greenlandic workers in general are dissatisfied with their jobs and have high job mobility due to an exit strategy. On the contrary, the results show that Greenlandic workers are highly satisfied with their jobs. The majority state that they are satisfied (90.7%) and feel they have a say about the job tasks they carry out (94.2%). The correlation between dissatisfied workers and job mobility might still occur in Greenland; however, data indicates this is not the sole explanation.

Another explanation for the high job mobility is contextual and relates to the fishing and hunting traditions in Greenland. Previous research points to the fact that many Greenlandic workers prefer to go fishing when the weather conditions allow it, even when they have work obligations (see Chapter 7 for discussion). Hence, it is not only dissatisfaction that pushes workers to shift jobs, but rather a desire to go hunting or fishing that frequently pulls workers out of employment. Workers going fishing or hunting instead of working creates challenges in the workplace because it translates into unsteady production capacities.

If it is not possible to change the workers' behaviour and make them stop fishing instead of working, the workplaces can adapt to the workers. Such a mentality is known from other fields (e.g., maternity and paternity leave helping families or wage subsidies, sheltered employment, and other precautions that aim to integrate persons with health issues or disabilities). It is worth investigating whether an employment agreement could integrate legal absences during peak hunting and fishing seasons. In Denmark, for example, an employee has the right to stay at home the first day a child is sick to take care of the child. The employee has to call the workplace as early as possible but can then stay at home still receiving salary. Likewise, a retention policy or strategy in Greenlandic workplaces could perhaps include the possibility for staying at home once or twice per year to allow for fishing or hunting. In that way, the employer can potentially limit or control employee absences and use fewer resources on recruitment.

5.6 Self-employment in Greenland

Because the Greenlandic economy has a long tradition of workers being self-employed, we find it relevant to include self-employment in our study. The self-employed in Greenland are distinguished from the rest of the workers. The self-employed are a relatively homogeneous group in Greenland; however, the self-employed include only 7.2% of all those employed.

Before we characterise the self-employed in Greenland, we address the question of why Greenlanders turn to self-employment.

There are many reasons for being self-employed. Nonetheless, in Greenland, nearly 60% of the self-employed point to either "flexible working hours" or "the opportunity to be self-employed occurred at a convenient time" as the most important reasons for becoming self-employed. Considering that Greenlandic workers have a high average number of working hours, the reason "flexible working hours" is interesting. Flexible working hours could address the need for adapting work hours to other daily activities.

Furthermore, our analysis shows that the motives for becoming self-employed typically are compound. The above-mentioned reasons are the most important ones. When asked for several reasons, nearly all self-employed respondents point to a number of reasons for their decisions to become self-employed. The motives for being self-employed are rather complex and include many aspects. Other than the above-mentioned reasons, the self-employed also maintain that they are carrying on a family business or that it is most common to be self-employed in the industry in which they work.

As mentioned before, the self-employed are quite a homogeneous group (cf. Table 5.3). Three quarters of all self-employed work in fishing, and the majority are men. Only 8% of all self-employed are women. This is a very surprising result. In 2019, the share of self-employed women represented 30.2% of all the self-employed in Denmark. Even though that is also a low share, the Greenlandic result is surprisingly low.

Almost 60% of the self-employed are older than 45 years. Another interesting result regarding the self-employed is that one third of all self-employed have worked at their present workplace for more than 20 years. They have been self-employed for a very long time. This is only the case for 7.5% of the non-self-employed workers, who have been working at their present workplace for far fewer years.

Table 5.3 Self-employed divided on gender and age, 2019 (in %)

Variable	Share of self-employed
Gender	
Women	8%
Men	92%
Age	
16–24 years old	7.8%
25–34 years old	13.7%
35–44 years old	19.9%
45–54 years old	35.3%
55–64 years old	23.5%

We also identify two educational backgrounds within the self-employed. One group can be defined as unskilled workers, who have primary school (45.1%) as their highest educational level, or skilled workers, with a vocational education (27.5%). The rest of the self-employed (27.4%) have other educational backgrounds including high school (2%), a bachelor's degree (3.9%), master's degree (2%), or none of the mentioned educational levels (19.6%). The latter include no education at all.

The study gives a clear picture of the self-employed as men older than 45 years of age, mostly unskilled or vocationally skilled, and having worked at the same place for a long time. The picture is consistent with traditional fishermen and points to the fact that this tradition still has significant impact on the lives of the Greenlandic self-employed (Bianco, 2019).

The result also points to potential within the self-employed. A further development could include women and other industries. Only a few women and industries are included in the area of self-employment in Greenland today, aside from the dominant fishing industry. Our results show that a fair share of women actually has a desire to become self-employed, and there is a gendered potential to increase the numbers of self-employed. In addition, the result shows that 74.5% of all self-employed work within fishing, hunting, and agriculture. Perhaps policy initiatives could provide other types of jobs that can be carried out by the self-employed. This could include industries like tourism, hospitality, or day care. By implementing an active industrial policy, the Government of Greenland can promote greater incentives for people to become self-employed. In the EU, initiatives such as Erasmus for Young Entrepreneurs aim to help young people become entrepreneurs (European Union, 2020), and the European Commission (2020) has created networks and online platforms in support of women entrepreneurs. Other initiatives could include campaigns informing citizens of opportunities to become self-employed. Our data indicates a potential for increasing the number of self-employed and that could call for a national strategy on entrepreneurship. Just like the European Union and European Commission target young people and women, a Greenlandic strategy could target specific industries.

5.7 Conclusions and discussion

In this chapter, we explored Greenland's labour supply by applying data from the first beta version of the Labour Force Survey conducted in Greenland. The lack of labour has been often spoken about and debated in Greenland. In general, there are two strategies to increase the labour supply (in addition to importing foreign labour): (a) increase employment of unemployed individuals and thereby include more people from outside the working force and (b) increase the amount of work done by already employed individuals.

In this chapter, we have quantitatively investigated the possibilities of two strategies in Greenland by using data from the first Labour Force Survey in the country. By applying the Labour Force Survey definition of unemployment, not working, actively seeking employment, and able to start work within two weeks, we find a relatively low unemployment rate and high employment in Greenland compared to selected countries. The low unemployment must be seen in the light of an economic upturn at the time of data collection in spring 2019; however, we might expect large regional differences due to the segmented labour market. Greenland has for the past few years seen a very positive economic development caused mainly by favourable prices on fish. The economic upturn is obviously connected to a high demand for labour creating low unemployment as a consequence.

The results also revealed that 19.3% of the population can be defined as inactive in Labour Force Survey terms. Of the inactive population, the data showed that some were unable to work due to health issues, and others were inactive due to ongoing, full-time education. The group who is inactive because of ongoing education is expected to gain employment in a foreseeable time. This is not necessarily the case for the other group included among the inactive. If they are to be employed, their individual health and social conditions have to change.

There is only a little (if any) potential for increasing the amount of work done by individuals already in employment. Greenlandic workers have long working hours, and only 10% work part-time. That is far less than the average of the European countries. On the contrary, Greenlandic workers are working at the same workplace for a very short period of time. Previous studies point to the fact that Greenlandic workers replace employment with fishing and hunting if the weather conditions allow it. Another explanation is that the fishing industry, in which many Greenlanders work, has high seasonal fluctuation, causing the dismissal and rehiring of workers.

Regarding the self-employed, the data shows a clear picture. The self-employed are primarily middle-aged males working solo by fishing from small open boats. Here we find the potential to increase the number of self-employed. Only a few women are self-employed, but many express a wish to become self-employed.

References

Abbott, M., & Ashenfelter, O. (1976). Labour supply, commodity demand and the allocation of time. *The Review of Economic Studies, 43*(3), 389–411.

Antropologerne. (2016). *Udfordringer og muligheder for ledige: En undersøgelse af lediges behov samt årsager til ikke at møde op til anvist arbejde* [Challenges and opportunities for the unemployed: A study of the needs of the unemployed and reasons for not showing up for assigned work]. Government of Greenland.

Bianco, N. (2019). Ender Grønlands økonomi og erhvervsudvikling i fisk? [Does Greenland's economy and business development ends in fish] *Politik, 22*(1). DOI: https://doi.org/10.7146/politik.v22i1.114839.

Danish Agency for Labour Market and Recruitment. (2020, May 31). *Bilag: Regional overvågning af situationen på arbejdsmarkedet, beskæftigelsesministeriets COVID-19 beredskab* [Appendix: Regional monitoring of the labor market situation, Ministry of Employment's COVID-19 contingency]. Status søndag den.

DST. (2019). *Arbejds-kraft-under-søgelsen, beskæftig-else.* https://www.dst.dk/da/Statistik/emner/arbejde-indkomst-og-formue/beskaeftigelse/arbejdskraftundersoegelsen

Eisenberg-Guyot, J., & Prins, S. J. (2019). Relational social class, self-rated health, and mortality in the United States. *International Journal of Health Services, 50*(1), 7–20.

European Commission. (2019). Employment and social developments in Europe, sustainable growth for all: Choices for the future of Social Europe. *Annual Report 2019.*

European Commission. (2020). *Women entrepreneurs.* https://ec.europa.eu/growth/smes/supporting-entrepreneurship/women-entrepreneurs_en

European Union. (2020). Erasmus *for young entrepreneurs.* https://www.erasmus-entrepreneurs.eu/

Eurostat. (2020a). *Eurostat.* https://appsso.eurostat.ec.europa.eu/nui/submitView-TableAction.do

Eurostat. (2020b). *Eurostat.* https://appsso.eurostat.ec.europa.eu/nui/show.do?dataset=lfsq_ewhais&lang=en

Eurostat. (2020c). *Eurostat.* https://ec.europa.eu/eurostat/web/lfs/methodology/main-concepts

Fein, O. (1995). The influence of social class on health status: American and British research on health inequalities. *Clinical Review, 10,* 577–586.

Government of Greenland. (2020). *Labour Market Report 2018–2019.* https://naalakkersuisut.gl/~/media/Nanoq/Files/Publications/Arbejdsmarked/DK/Arbejdsmarkedsredeg%c3%b8relsen%202018-2019%20DK.pdf

Groves, R. M., Fowler, F. J., Couper, M. P., Lepkowski, J. M., Singer, E., & Tourangean, R. (2009). *Survey Methodology.* Second edition. Wiley.

Høgedahl, L., & Krogh, C. (2020). *Den Grønlandske Arbejdskraftundersøgelse* [Greenlandic Labour Force Survey]. Aalborg Universitet.

Jahoda, M., Lazersfeld, P. F., & Zeisel, H. (1971). *Marienthal: The sociography of an unemployed community.* Aldine.

Kenny, K., Whittle, A., & Willmott, H. (2011). *Understanding identity & organizations.* SAGE.

Kristensen, N., & Westergård-Nielsen, N. (2004). Does low job satisfaction lead to job mobility? *IZA Discussion Paper Number 1026.*

Leidner, R. (2006). Identity and work. In M. Korczynski (Ed.), *Social theory at work*. Oxford University Press, 424–463.

Mackenbach, J. P. (2006). Health inequality: Europe in profile. University Medical Center Rotterdam.

Madsen, P. K. (2004). The Danish model of 'flexicurity': Experiences and lessons. *Transfer: European Review of Labour and Research, 10*(2), 187–207.

Marquardt, O. (2005), Opkomsten af en grønlandsk arbejderklasse [The rise of a Greenlandic working class]. In A. Carlsen (Ed.), *Arbejdsmarkedet i Grønland – fortid, nutid og fremtid, 2005* [The labour market in Greenland - past, present and future, 2005] (pp. 176–188). Ilisimatusarfik.

Moscarini, G., & Thomsson, K. (2008). Occupational and job mobility in the US. *The Scandinavian Journal of Economics, 109*(4), 807–836.

OECD. (2018). *OECD.Stat.* https://stats.oecd.org/

OECD. (2019). Health for everyone? Social inequalities in health and health systems. OECD Health Policy Studies. OECD Publishing. https://doi.org/10.1787/3c8385d0-en

Petersen, H. (2005) Etnicitet og arbejdsmarked i det 20. o 21. århundrede [Ethnicity and labor market in the 20th and 21st century]. In Aksel V. Carlsen (Ed.), *Arbejdsmarkedet i Grønland – fortid, nutid og fremtid* [The labour market in Greenland - past, present and future, 2005] (pp. 125–140). Ilisimatusarfik.

Prattis, J. I., & Chartrand, J. P. (1990). The cultural division of labour in the Canadian North: A statistical study of the Inuit. *Canadian Review of Sociology/ Revue canadienne de sociologie, 27*(1), 49–73.

Price, R. H., Friedland, D. S., & Vinokur, A. D. (1998). Job loss: Hard times and eroded identity. In J. H. Harvey (Ed.), *Perspectives on loss: A sourcebook* (pp. 303–316). Taylor & Francis.

Poppel, M. (2005). Barrierer for grønlandske mænd på arbejdsmarkedet [Barriers for Greenlandic men in the labor market]. In A. V. Carlsen (Ed.), *Arbejdsmarkedet i Grønland – fortid, nutid og fremtid* [The labour market in Greenland - past, present and future, 2005] (pp. 125–140). Ilisimatusarfik.

Rosenfeld, R. A. (1992). Job mobility and career processes. *Annual Review of Sociology, 18*, 39–61.

Saammaateqatigiinnissamut Isumalioqatigiissitaq (2017) *Det Grønlandske Sprog I Dag* [The Greenlandic Language Today]. https://naalakkersuisut.gl/~/media/ Nanoq/Files/AttachedFiles/Forsoningskommission/Oqaatsitta inissisimaneratDK.pdf.

Thorleifsen, D. (2005). Arbejdskraftens sammensætning og spørgsmålet om planlægning og uddannelse i 1950'erne [The composition of the workforce and the issue of planning and education in the 1950s]. In A. Carlsen (Ed.), *Arbejdsmarkedet i Grønland – fortid, nutid og fremtid* [The labour market in Greenland - past, present and future, 2005] (pp. 189–202). Ilisimatusarfik.

6 Social security and barriers to labour market participation

Niels Jørgen Mau Pedersen

Abstract

This chapter analyses the public assistance scheme in Greenland that being the economic safety net of last resort for part of the population. Public assistance applies relatively low rates, yet they are tax free and measured out with some discretion and compensation of certain family fixed cost. Based on a large evaluation, we question if the system works effectively? First, we look into the attitude towards the recipients. We find the system to be paternalistic with a low level of transparency. Instead of empowering individuals, this may imply a risk of discouraging them from effective job search and increase of qualifications. Second, we consider economic incentives for individuals getting a job and income, i.e. compound marginal tax rates. Here we discover possible high marginal compound tax rates, e.g. for part-time workers. Third, we address incentives at the municipal task level. We also look into the administration or governance of municipalities and find some tasks rather complex and possibly costly. Summing up, we deliver a few recommendations for developing the public assistance system.

6.1 Background, scope, and program

In the Greenland labour market, we see significant differences in unemployment levels among various parts of the working population, lack of certain kinds of labour, and a strong demand for both private and public sector employment. Consequently, a wish to increase the level of employment in the workforce is on Greenland's political agenda. Before the COVID-19 crisis, we saw strong GDP growth and a shortage of labour in Greenland (Christensen, 2018; Chapter 2 in this book). The COVID-19 crisis has caused a decrease in employment in Greenland, but there is still a desire to increase it again in the future, certainly among skilled labourers (Department of Finance, 2019, p. 50).

Simultaneously with demand for – and lack of – labour, we have seen a substantial portion of unemployed persons receiving public assistance (*offentlig hjælp*) for several years. In reality, many of those persons are not available for the job market. This fact and other aspects raise questions about how up to date and effective the existing public assistance system is.

In 2016, the Department of Social Affairs, Families, Gender Equality, and Legal Affairs in Greenland asked VIVE, the Danish Center for Social Science Research, to investigate the system of public assistance in Greenland, and in 2017–2019, VIVE conducted a comprehensive analysis of the whole system (VIVE report by Mau Pedersen et al., 2019b). According to the finance bill proposal for 2020 and 2021 (Grønlands Selvstyre, 2019, p. 143, 2020, p. 148), there are plans to reform the public assistance system, and the coalition agreement of the ruling government (Naalakkersuisut, 2020a) also foresees a possible reform of labour market regulation. The headlines of the reform underway are according to the finance bill higher level of employment, improving incentives for employment, clear rights and obligations, less discretion in measuring out public assistance, and releasing administrative resources.

Against this background, the aim of this chapter is to investigate and analyse the barriers to labour market participation in Greenland, especially for persons that already receive or may in the future receive public assistance – and taking into consideration the current system of public assistance.

The analysis will make use of the findings in the VIVE report (Mau Pedersen et al., 2019b) mentioned above, also considering other relevant studies. Aside from the VIVE study, however, we have not identified other overall studies on the public assistance system, but reports from, for example, Skatte-og Velfærdskommissionen (2011) and Greenland Economic Council (Grønlands Økonomiske Råd, 2020) also address relevant topics.

First, we briefly discuss the Greenlandic economy in relation to the public assistance system. We also include some historical facts on the background of the assistance system.

Next, we analyse the assistance system from different angles on the question of possible barriers to labour market participation. There exist many such angles or dimensions of this issue, but we will concentrate on three types of barriers, which we deal with separately.

First, we consider the system's attitude towards public assistance recipients. In other words, we investigate the possible barriers stemming from this attitude or the way the system meets its "customers". This includes the question of how unemployed persons handle their own occupational and economic situations.

Second, we focus on the individual incentives for seeking a job (or beginning an education) to improve participation in the labour market. This includes calculation of the extra net income resulting from the unemployed person gaining income from business activity (e.g. wage income), but on the other hand, they must foresee a reduction of public assistance and other income-related benefits such as housing assistance and possibly a tax bill.

Third, we discuss the governance dimension of the public sector tasks with labour market participation (i.e. the administrative and organisational aspects). This dimension contains, as one element, the question of how to establish a financial system of the local governments that supports the Active Labour Market Policy (ALMP) of the municipalities. Moreover, we look into the question of how to create a system and organise a public assistance scheme that permits effective administration of the system.

Finally, we conclude with a short discussion of recommendations to change or correct the existing system of public assistance deduced from the analysis.

6.2 Public assistance and the Greenlandic economy – development and history

Public assistance is an important part of the social security system of Greenland when it comes to compensation for a loss – or lack – of private income. The public assistance system is the last-resort element of social security. There seems to be no maximum period during which a person can receive assistance.

Another element of the social security system is unemployment benefits (*arbejdsmarkedsydelse*). The public sector provides these unemployment benefits in situations of unemployment or sickness when the person beforehand has had a labour income (i.e. wage income). The scheme provides these benefits for a maximum of 13 weeks (not taking into account benefits in the case of sickness). Among other elements of the social security system are rehabilitation benefits and early retirement pension, wage support of certain jobs (*fleksjob*), maternity benefits, and education benefits (*uddannelsesydelse*). Here, we consider especially the public assistance system that constitutes the social security economic safety net of last resort for people in an economically needy situation. As indicated, this may be a short- or long-term situation but also includes what might be termed emergency economic support and support (or compensation) of certain fixed cost to the person or family.

As a starting point, public assistance is not taxable, unlike the other benefits of social security mentioned above. In principle, the municipal authority provides the assistance on a discretionary basis, considering the previous living conditions, the actual necessary expenditures and the local conditions, possible maintenance obligations, etc. The person may also receive supplementary public economic support from child benefits, reduced day care payment, alimony, and/or housing subsidies.

Of course, there are analogies and similarities with parts of the Danish social security system (*kontanthjælp*, cash benefits). However, it might be more interesting to note that there seems to be a common origin historically, followed by a deviant development.

Historically, Greenland's public assistance back in the 1980s, according to the regulation of 1982, had strong similarities to the former Danish social security system of that time (*bistandshjælp*; cf. the basic and seminal Danish reform of social security in 1976). Around 1980, the Danish social security system calculated payments on a discretionary basis for individuals, considering former living conditions. However, whereas the Danish system in 1987, considering the interesting socio-economic "social security experiment" (Koch & Rørbeck, 1985), replaced the discretionary benefits with legally fixed benefits in most situations, the discretionary assistance benefits still exist, at least in principle, in Greenland (see the 2006 regulation; Lovgivning.gl, 2006). This is the case, although the assistance rates in Greenland are in practice more or less fixed *within* the single municipalities, towns, or villages. Notably, already in the 1990s, the Greenland Social Reform Commission (Socialreformkommissionen, 1997, p. 22f) criticised this variation in rates between municipalities and between towns and villages.

It is noteworthy that the local conditions may influence the size of public assistance benefits in Greenland. Likely, this mirrors the fact that contributions from personal resources in connection with nature (fishing, hunting, etc.) may formerly, in the society of Greenland, have played a significant role in families' living conditions, depending on the conditions of the given locality (Mau Pedersen et al., 2019b, p. 21).

Regarding the volume of the public assistance (i.e. the total societal costs of the scheme), the economic activity level causes significant variations from year to year. The number of persons receiving public assistance peaked around the years 2013/2014 at 8,000 persons, but thereafter it is markedly reduced, to reach about 4,500 in 2019 (see Figure 6.1). Correspondingly, expenditures in 2014 reduced from 162 million DKK annually in running price level to a little less than 76 million DKK in 2019. As already mentioned, this may go hand-in-hand with a shortage of labour in certain professions and localities, even though one might expect the COVID-19 crisis to imply a certain economic downturn (Grønlands Økonomiske Råd, 2020, which indicates a GDP reduction of 1.5%).

It could be mentioned here – perhaps a bit cursorily – that VIVE, in all its investigations carried out for Naalakkersuisut and the Department of Social Affairs in recent years, has encountered elements of labour shortage, especially skilled labour from social workers (Dahl et al., 2020, p. 9; Holt et al., 2019, p. 19; Mau Pedersen et al., 2016, p. 43). Another observation from these studies is that on the one hand, public assistance recipients often lack relevant education, and on the other hand, only a limited portion of social workers in social departments and Majoriaq centres (*jobcentre*) have a relevant vocational or academic education. Therefore, lack of relevant educational background in itself may constitute a barrier to labour market participation (see also Chapter 7). From the 2018 VIVE report (Mau Pedersen et al., 2016, pp. 54–65) and a special statistical run from Statistics Greenland, we know

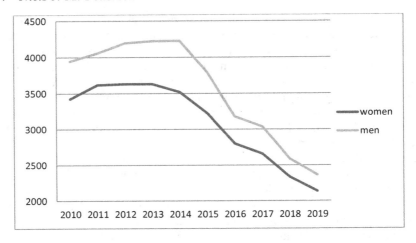

Figure 6.1 Development in number of persons receiving public assistance divided according to gender, 2010–2019.

Note: N = 4.498–7.854. Includes only persons of the age of 18–64 years.

Source: Grønlands Statistik, SOD004.

that 81% of public assistance recipients in 2016 only had a diploma from ordinary public school as their educational background. The VIVE report discusses several aspects of this situation, including instructions tailored to public assistance recipients, the so-called campus model, language courses, so-called AMA and PKU courses, etc. However, we do not discuss these issues more in depth in this chapter (please see Chapter 4).

According to Mau Pedersen et al. (2019b, p. 47ff), the share of the population to receive public assistance within a year varies significantly between the municipalities. We observe the biggest share of the population (age 18–64 years) in the Kujalleq Municipality, with as much as 23.3% (2017). Furthermore, we note that among the recipients of public assistance, younger persons (age 18–29) in particular experienced a significant reduction in number from 2010 to 2017. However, currently, those younger age groups still are overrepresented among the recipients compared with elderly age groups.

6.3 Public assistance system: the attitude towards recipients

The design of the system of public assistance in Greenland and the administrative way of handling the social matters reflects what we might denote as the attitude towards the potential and actual recipients. This approach can have consequences for citizens' experience of the situation (i.e. the personal feeling, or understanding, of responsibility for one's own life situation and especially personal economy and one's ability to influence the success of labour market participation). In other words, if public assistance system causes individuals to feel less responsible for the development of their own situation than would otherwise be the case, this might constitute a barrier to job search, acquiring competencies, etc.

In a literary context, we might term this a lack of bureaucratic self-efficacy (i.e. beliefs about one's capability to perform successfully in the labour market and be employed; see Bandura, 2006, regarding social cognitive theory), a concept that might be linked to the concept of empowerment (see below).

In sum, the significance of these conditions could have implications for the individual behaviour in relation to labour market activity. The VIVE study (Mau Pedersen et al., 2019b, p. 51ff) does not include separate independent studies of these professional issues. However, there exists a relevant Danish study referring to the research area (i.e. the major Danish project of employment from 2012; Beskæftigelsesindikatorprojektet, 2012), which includes six themes with relevance for individual job readiness:

- *Identity and coping*, i.e. self-perception and confidence in being able to meet personal goals.
- *Social competencies*, i.e. ability to engage in constructive relationships with others.
- *Personal competencies*, i.e. characteristics exerting influence on the ability to complete daily chores.
- *Physical and mental health situation*, i.e. health conditions relevant to being able to perform a job.
- *Job search*, i.e. job search behaviour using informal networks or formal channels.
- *Professional competencies*, i.e. total knowledge and skills for relevant jobs.

Each of these items really deserves a separate discussion. However, we focus here on the combination of identity, coping, and personal competencies, which, taken together, may be described by or related to the concept of *empowerment* in contrast to *learned helplessness* (Campbell & Martinko, 1998; Caswell, 2005; STAR, 2014).

Currently, the public assistance system in Greenland in this respect has several characteristics that may negatively influence the recipients' sense of empowerment in relation to the labour market. The unemployed person receiving public assistance could thus be at risk of having low levels of self-esteem and self-confidence. Similarly, identity and confidence in one's own abilities on the one hand, and on the other hand, the ability to organise and solve a problem (e.g. participation in the labour market) are closely connected (Beskæftigelsesindikatorprojektet, 2012, p. 46).

As mentioned above, one characteristic of the public assistance system is a discretionary approach to measuring out the size of the benefits, considering former living conditions and social circumstances. Even though the municipal authorities calculate part of the assistance, known as the disposable amount (*rådighedsbeløbet*), using fixed rates, those rates vary across authorities in accordance with historical factors and the local practice, depending on family and demography. In addition, the household receives a supplementary child amount according to the number of children that the family has to take care of. Moreover, an important part of the public assistance is what is often termed sensible expenditures (*rimelige udgifter*) for housing, user payments, etc. (Lovgivning.gl, 2006). This might include expenditures for public utilities such as electricity, water and heating, day care payments, sensible expenditures for glasses and necessary expenditures for clothing, support of certain mortgages, and (for some authorities) professional contingents. In total, two families that are rather similar might nevertheless receive public assistance that differs significantly in amount.

On this basis, we assess the transparency of the public assistance system to be rather low for many potential recipients. Furthermore, the lack of taxation for the typical recipient inhibits comparability with other kinds of income. Expressed in a thought-provoking manner in the VIVE report (Mau Pedersen et al., 2019b, pp. 90–91), the VIVE assessment is (own translation) that "in many situations, the recipients understand the public assistance system as an impenetrable system and therefore often have to ask the caseworker to make a calculation about the eligibility for public help and the possible size of it".

In line with this, the approach to and confidence in the recipients' handling of personal economy often seem to be rather sceptical or negative when it comes to people's use of the amounts paid out from the public assistance system. Here, the assessment for two out of three visited and investigated municipalities was that most public assistance recipients were not able to handle the assistance payments responsibly. Consequently, the public authorities often incur the fixed expenses of the individual and pay a residual amount to cover food, etc. For one of the municipalities involved in the investigation, the municipality incurs the fixed expenses for 90% of the recipients, leaving only 10% of them to administer the total public assistance amounts.

From the perspective of some of the caseworkers of the municipalities, the disposable amount is relatively small, which in itself may be a constraint for citizens receiving public assistance to manage their own economy (Mau Pedersen et al., 2019b, p. 33, 60). One of the municipalities, however, has changed its practice and seems to be satisfied with it. The change implies that instead of letting the municipal authority and its caseworkers pay the fixed costs for the needy citizens, the citizens now receive a disposable amount of money but are then themselves responsible for paying the bills.

In conclusion, even though the purpose of the public assistance legislation is possibly to make the recipients more accountable for their own economy, this system may seem rather discretionary and difficult to comprehend. We could perhaps characterise it as a rather paternalistic system (cf. literature on welfare dependency, e.g. Freud, 2007). If in reality the system likely does not to a sufficient degree encourage individual accountability, but the opposite, in some cases, it will constitute a hindrance to labour market participation.

Even though the way of treating recipients described above without doubt has its justifications, the above revealed characteristics of the scheme are nevertheless remarkable. The purpose of the system is to increase the economic security of the target group by protecting people. However, lack of transparency could possibly reduce or undermine the security.

A nuanced observation of the public assistance system is that it has been an obligation since the 2006 regulation to put up an "action plan" for the single individuals to help them handle their situation. In practice, the development of these action plans and subsequently following up varies between authorities.

6.4 Public assistance and individual incentives to participate in the labour market

One standard assumption among economists about individual labour market behaviour is that economic incentives have an impact on the individual's job search activity and motivation to acquire new competencies (Mau Pedersen et al., 2019a, p. 24). We later discuss, under the heading of "governance", incentives as an instrument of control within the public sector itself. In this section, we concentrate on the first-mentioned kind of incentives relevant for the individuals[1] For a discussion of measurement, impact and theory of incentives for municipalities, see Mau Pedersen et al (2019a, chap. 3).

The incentives perceived by the individual primarily depend on what might be termed the compound marginal tax rate when a person obtains employment and earns a private income. This includes both ordinary income tax from the private business income and the possible reduction of public assistance and other public benefits (e.g. housing benefits) when the person in question improves his or her own economy by earning private income. This issue connects with interaction problems among the systems of taxes and public benefits and incentives to work. As an alternative to calculation of compound marginal tax rates, we can set up calculations of the total net disposable income for different types of individuals and households who possibly receive public assistance but might also receive a private business (wage) income of a certain amount. For the analysis, the investigators (Mau Pedersen et al., 2019b, pp. 36–46) developed a separate calculation model to illuminate those matters. This model takes into account conditions for public assistance, housing expenditures, day care fees, taxation rules, private income under certain assumptions, etc.

The calculations for various types of households take into account the conditions in 2017/2018, minimum wage for the members of the Greenland trade union "SIK", and number of children in the households and their ages.

It is evident, as already indicated, that public assistance depends rather heavily on the number of small children in the family because of the supplemental assistance for children taken care of in the family. This probably mirrors the fact that the living conditions of families with children are a politically high-valued priority in Greenlandic society. Aside from the number of children, the level of public assistance depends on the number of adults in the household, personal private income, the private income of a possible cohabitant, and the fixed expenses of the family household. The final result of the calculations is called the "net degree of compensation", which is a ratio (%) measured as the total calculated disposable income for the family in consideration of receiving public assistance in relation to the disposable income of a family with both adult members earning a wage income (i.e. minimum wage of SIK members). The calculations apply to singles and couples, respectively, where for couples, both or only one person receives public assistance. The calculations include wage income as well as public assistance, unemployment benefits, and education benefits.

In general, the model calculations reveal dependency of the net degree of compensation on number of children in the household. For couples with three children, where both adults receive public assistance, there is a negative payoff for the family if one of the adults gets a job paid with the SIK minimum wage, since the net degree of compensation is reduced from 77% to 73%. Furthermore, based on the calculations, the problem of interaction between benefit schemes and tax rules seems to be most distinct for low-paid jobs, including part-time jobs.

The issue of possible interaction problems and high compound marginal taxes for individuals acquiring part-time jobs seems to be related to the way municipalities offset public assistance. Here, the municipal authority can choose between two possible offsetting rules:

1   Number of days with a wage income multiplied by a calculated rate per day for public assistance.
2   The total earned amount of wage income.

Even though the first-mentioned possibility referring to the Tax and Welfare Commission (Lovgivning.gl, 2006; Skatte- og Velfærdskommissionen, 2011) seems to have been introduced to improve the incentives for individuals to engage in part-time jobs, the municipalities, according to interviews, seem to only make use of the second option (Mau Pedersen et al., 2019b, p. 31). This means an offset by 100% of wage income. This 100% taxation could constitute a serious hindrance to part-time jobs being attractive to individuals receiving public assistance.

The problem of incentives is also relevant when it comes to short-term jobs. In this sense, it may, however, also be noted that the calculation of so-called emergency public assistance considers only the private income for the latest three months – and not, for example, the latest full year (Mau

Pedersen et al., 2019b, p. 33, 46). This factor possibly, depending on the circumstances, makes short-term work opportunities more favourable for the individual than would otherwise be the case.

Furthermore, the model calculations reveal that the unemployment benefits are at a higher level of compensation than the calculated public assistance, despite the benefits being taxable. On the contrary, the net degree of compensation is generally lower for households living off educational benefit schemes than off public assistance.

Moreover, the fact that the income of a spouse or co-supplier is offset fully in public assistance of the other spouse involves a risk that some spouses or cohabitants do not find it economically tempting to actively search for a job. This phenomenon is, however, also well known in Denmark.

To conclude, in the existing system, we find certain problems of interaction and weak incentives to participate in job search.

6.5 Public assistance: efficiency of the administration and public use of resources – (governance)

In this section, we deal with the public sector organisation and task performance – what might be termed "governance". Our agenda in this respect is how this governance and effectiveness may depend on *first,* the system of municipal finance, and *second*, the character of the tasks that the municipality has to perform.

6.5.1 The financial system of municipalities in Greenland

In continuation of the previous section about incentives for individuals, it is relevant to note that incentives also play a role within the public sector itself, including the municipalities. These economic incentives for the municipalities to work effectively with the labour market issues depend, among other factors, on the design of the public sector financial system, whereby the central level of the public sector (i.e. the Central Government) transfers financial means to the municipality, assuming that the municipalities do not have to finance their expenditures for public assistance and other labour market activities solely through their own tax income or similar sources of income. Furthermore, this particularly implies the central-level reimbursement (*refusion*) system of the municipal expenditures for public assistance (i.e. the percentage rate that the central level uses when calculating the size of the earmarked reimbursement of municipal expenses). Normally, we see this system of reimbursement, in Greenland as well as in other countries, together with the equalisation system between the municipalities, because equalisation concerns the net expenditures for municipalities when the reimbursements payments have been deducted (Mau Pedersen, 2017).

The issue of equalisation as such, and how to evaluate the equalisation system, is not a subject of this analysis. We note that Greenland for several years and also both 2020 and next year in 2021 applies a temporary equalisation scheme between the municipalities. The finance bill stipulates the exact transfers to and from the five municipalities in Greenland (i.e. a negative amount/contribution for Kommuneqarfik Sermersooq and positive amounts/grants for the other four municipalities; Grønlands Selvstyre, 2020, p. 240; Lovgivning.gl, 2019).

From an overall fiscal federalism perspective, three separate kinds of considerations are relevant to discuss when determining the size of the central level's reimbursement of municipal expenditures for income transfers (i.e. the expenditures that the municipality would otherwise have to be financed by taxes and/or block grants):[2]

1   The consideration of *equalisation*
2   The consideration of *accountability*
3   The consideration of giving *incentives*

Depending on the weighting of those considerations, different systems may seem appropriate.

The main priority of the first consideration, *equalisation*, is to enable the municipalities to have equal financial opportunities to meet the expenses, including the expenditures for public assistance. These expenditures seem to vary considerably between the municipalities. The explanation for part of this variation presumably is variation in rates, which the municipalities have mainly chosen for themselves. Another factor is the share of the population that receives public assistance. This share varies across municipalities (e.g. within the year 2017, from 13.1% for the Sermersooq Municipality to 23.3% for the Kujalleq Municipality).

If the central government level wants to obtain full equalisation of the relevant expenditures, the reimbursement rate should be at 100%. Alternatively, equalising the expenditure needs of the area may be based on so-called objective needs criteria.[3]

The drawbacks of a 100% reimbursement system would possibly be a weakening of the interest of municipalities in keeping public expenditures as low as possible. This leads to the next consideration: accountability.

The consideration of *accountability* stresses the local authorities' responsibility for managing the tasks of the area appropriately and limiting expenses. We can express the consideration via the following slogan-like sentence: "Accountability and competency should go together!" (on the issue of accountability chain, see Kim et al., 2013). In other words, the level of decision-making towards the citizens (i.e. price for a certain municipal activity) should coincide with the actual "price" of their decisions. If the central level (the Central Government) reimburses the expenditures, it could blur the perception of the price because another public sector level pays part of the expenses. The Tax and Welfare Commission (Skatte- og Velfærdskommissonen, 2011) puts it this way (in translation): "In general, reimbursement systems are inappropriate when they encourage use of public services to an extent that there is really no need for" (p. 444).

Obviously, the consideration of accountability is most relevant when the municipalities have some degree of freedom in handling the expenditure area in question.

It seems that this consideration through the years has gained ground in the Greenland public sector. Thus, for both the municipal expenditures for public assistance and for unemployment benefits, there are no reimbursements (i.e. a reimbursement rate of 0 [nil]).

The third consideration is the wish to offer *incentives* for a certain municipal behaviour.

This consideration applies to a situation where the central level of public decisions (i.e. the Central Government) prefers a certain behaviour of the local governments (municipalities). If the Central Government finds it unlikely that the municipalities will perform this behaviour by themselves, the government may encourage the municipalities to behave in that way via financial incentives. Here, the reimbursement system is an instrument for the Central Government to establish such incentives for the municipalities. The challenge is how to design the reimbursement system to deliver the incentives appropriately.

First, the Central Government has to decide what kind of municipal behaviour it wants to encourage. The government would probably like to increase the level of employment in general, and especially for recipients of public assistance benefits. This might be via the general attitude towards the recipients of public assistance from the municipalities, as discussed in previous sections, or via the administrative organisation (e.g. in Majoriaqs; Holt et al., 2019; Chapter 6). Alternatively, it could be through the use of different remedies and instruments of employment policy, such as training and activating programs, educational efforts, employment subsidies, etc. (i.e. what we term the ALMP). Here there exists a vast amount of literature and international studies on the effects of using such instruments (e.g. Card et al., 2010, 2017; Lindsey & Mailand, 2004).

The Danish Labour Commission (Arbejdsmarkedskommissionen, 2009) presents the idea of designing reimbursement systems regarding incentives for the ALMP. Subsequently, in 2016, the Danish Government implemented a reform of the reimbursement system for municipal labour market income transfers. Regarding the system of public assistance in Greenland, we observe that a significant portion of the recipients has been receiving public support for a considerably long time (i.e. 2–3 years; Naalakkersuisut, 2016, p. 6).

One way to encourage rapid municipal action for unemployed individuals would be to establish a relatively high or generous reimbursement rate for expenditures concerning the unemployed with a short period of public support, but diminishing the reimbursement rate or even abolishing it for persons receiving public benefits for a long period. However, letting the reimbursement rate be dependent on the length of the period of public support naturally demands robust data and a rather sophisticated system of payments from the central to the local level. Moreover, if social and psychological barriers are the main factors explaining the long periods of public support, we might question the effects.

Figure 6.2 shows five models of principal reimbursement systems including income transfer schemes in Greenland. As can be seen, the models differ from each other by the design of reimbursement rates, hereby giving different weights to the three previous considerations.

The first part of Figure 6.2(a) represents the existing system for the types of income transfers considered: unemployment benefits (*arbejdsmarkedsydelse*),

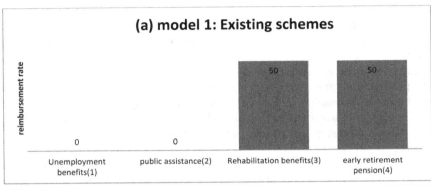

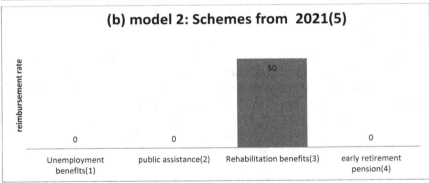

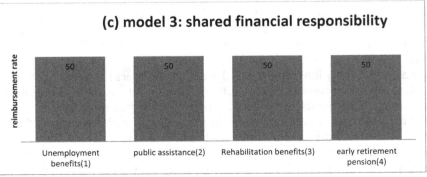

Figure 6.2 (Continued)

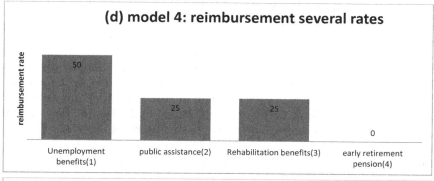

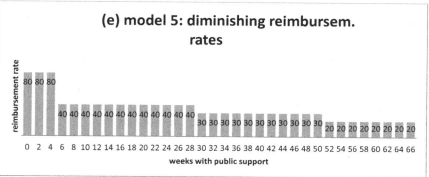

Figure 6.2 Different models of models of reimbursement schemes including various types of income transfers existing in Greenland.

Note (5): Regulation – a 5-year period of transition.

Source: Mau Pedersen et al (2019).

public assistance (*offentlig hjælp*), rehabilitation benefits (*revalideringsydelse*), and early retirement pension (*førtidspension*). Only for the two last mentioned benefit schemes does a partial reimbursement of municipal expenditures apply.

The second part of Figure 6.2(b), shows the future system where reimbursement of early retirement expenditures are abolished with a 5-year-transition period (Lovgivning.gl, 2020, para. 40, 44). Presumably, the consideration of local accountability has been given weight.

In the third part of Figure 6.2(c), we assume that the central level (the Central Government) and the municipal finances share expenditures on equally, with a reimbursement rate of 50%. Such a system prioritises the equalisation considerations but still has accountability in mind.

The system shown in the fourth part of Figure 6.2(d), financially encourages the municipalities primarily to avoid long-term public support as much as possible (e.g. public assistance in some instances and certainly early retirement, which often may be a permanent transfer). On the other hand, a high reimbursement rate applies to unemployment benefit expenditures, and the scheme hereby targets short-term periods of public support.

The principle of incentives via the reimbursement scheme is refined in the last part of Figure 6.2(e), with the reimbursement rate gradually diminishing according to the length of the period of income transfer to the individual. Denmark implemented such a system in 2016, and VIVE has evaluated the reform. VIVE (Mau Pedersen et al., 2019a) finds that the new system of reimbursement to some extent works in the direction of shorter support periods, but also involves some so-called cream skimming that favours certain unemployed groups (i.e. those that most easily obtain employment).

6.6 The tasks of the municipalities and administrative efficiency

The public assistance system legislation determines under what conditions the local authorities solve the related administrative tasks. This includes determining individuals' eligibility to receive assistance and measuring out the exact amounts. The design and possible complexity of the tasks is decisive for the working conditions of the local authorities.

With the existing scheme including some discretion when measuring out the size of the public assistance and finding a suitable level of compensating fixed expenses, the authorities' tasks may be rather extensive and perhaps complicated. Mau Pedersen et al. (2019b, p. 25, 89) found that the tasks could be comparatively resource demanding (i.e. the presence of a significant number of municipal caseworkers in the social administration offices).

This seems to be the case, while at the same time the municipalities highlight the challenges of recruiting the necessary number of social workers to the field of public assistance. The authorities involved express the dilemma by pointing out that more resources necessary for case management and administration means fewer resources for active policy in the labour market area, and vice versa.

In this context, it may seem remarkable that this was exactly the conclusion or statement from the Danish social security experiment from the 1980s (see the preceding section about development and history). The results from this experiment were part of the research background when the Danish system of that time changed the discretionary way of deciding the amount of individual social security payments to a system where the Central Government regulates most of the rates.

6.7 Concluding remarks and possible recommendations

A number of conclusions and recommendations derive from the preceding analysis presented in this chapter. There is a broad range of statements in this respect, reaching from aspects of the equal treatment of citizens across local authorities to administrative procedures, matters of transparency for potential public assistance recipients, interaction problems between tax regulation and income transfer schemes, and the organisational framework. Also relevant are issues concerning possible lack of competencies among the unemployed persons and perhaps further education for them. These final subjects, however, are often close to the activity of the municipal job centres (Majoriaq), which we refer to in Chapter 4.

In this context, we will concentrate on the elements that deal with potential barriers to labour market participation connected with the existing public assistance system. We address this subject by distinguishing between three different approaches to the issue of unemployed individuals receiving public assistance.

Some of the experiences and analysis, however, seem to end up with attention points and recommendations that are rather similar. This is true, on the one hand, for the seeming lack of support from the public assistance system to empower the supported individuals. The reason for that is, among other factors, the discretionary element of measuring out the disposable public assistance amount and the way of compensating certain fixed expenses. On the other hand, this system also implies an administrative burden and complexity of the local authorities' tasks that is rather costly in terms of local administrative resources. Thus, for both reasons, the first systemic change to consider might be to change it to a system based mainly on fixed rates for the disposable amount of public assistance instead of the partly discretionary allocation of public assistance. The second consideration might be a more simplified system of compensating fixed expenses for the recipients (e.g. by including a bigger portion of compensation for fixed expenses in the disposable amount, which would in itself underline the individual's own responsibility of private economy). A modification of the existing system of public assistance along these lines would likely make the system more transparent and simple, both for the individuals and for the local administrations. Additionally, the possibility to transfer (part of) the public assistance to a so-called closed account (payment account) owned by the recipient, from which only bills for certain expenses may be paid, should be mentioned. This closed-account system could possibly be a substitute for the existing system, where the local authority often not only compensates certain fixed expenses on behalf of the individual but also takes care of the payment process itself.

It would be in accordance with efforts to increase the sense of empowerment among the recipients to make the public assistance taxable in general. Hereby, the system of public assistance and the unemployment benefits scheme would be proportional in this respect.

Regarding possible barriers to entry into the labour market, high marginal compound taxes and interaction problems between the tax system and income transfer schemes may pose a problem. However, no easy solutions to such problems seem obvious. Of course, a general reduction of the level of public assistance and/or supplementary child assistance would in isolation ease or dampen the problems but would on the other hand jeopardise the system's ability to play the role of economic safety net of last resort for individuals. Focusing on improving the incentive to take a part-time job, it would be relevant to encourage the local authorities to apply the model of offsetting in public assistance that makes use of hours worked instead of earned income, as the regulation already allows today. Possibly, other kinds of deductions in earned income before offsetting public assistance may be relevant to reduce the problem of 100% marginal compound taxation.

Finally, the financial system of the municipalities may be subject to a revision. As for the marginal tax rates, however, an optimal change to the existing system is not obvious. To establish incentives to encourage ALMP, a change in the reimbursement schemes with reimbursement rates diminishing along with the length of the period of individual public support might be worth considering. However, it would be important to know more about the effect of the incentives (i.e. the so-called elasticities) before implementing such a system (Mau Pedersen et al., 2019a). Moreover, the risk of a more complex financial system and need for robust and precise data may constitute an obstacle here. Perhaps it would be more tempting first to further investigate the other elements of the municipal financial system. First and foremost, this comprises the equalisation system between the municipalities, which today seems rather simplistic.

Notes

1. For a discussion of measurement, impact, and theory of incentives in the labour market, see Mau Pedersen et al. (2019a, pp. 25–29) and references here.
2. This is a classical so-called fiscal federalism issue; see, e.g., Mau Pedersen et al. (2019a, Chapter 2).
3. That is, an equalisation system in Greenland along the lines that the government is planning for (Naalakkersuisut, 2020b).

References

Arbejdsmarkedskommissionen. (2009). *Velfærd kræver arbejde [Welfare demands work]*. Cph.
Bandura, A. (2006). Towards a psychology of human agency. *Perspectives on Psychological Science, 1*(2), 164–180.
Beskæftigelsesindikatorprojektet. (2012). *Litteratur-review ifm. BeskæftigelsesindikatorProjektet. [Literature review from employment indicator project]*. Væksthusets Forskningscenter.
Campbell, C. R., & Martinko, M. J. (1998). An integrative attributional perspective of empowerment and learned helplessness: A multimethod field study. *Journal of Management, 24*(2), 173–200.
Card, D., Kluve, J., & Weber, A. (2010). Active labour market policy evaluations: A meta analysis. *The Economic Journal, 120*(November), F452–F477.

Card, D., Kluve, J., & Weber, A. (2017). *What works? A meta analysis of recent labour market program evaluations* (NER Working Paper no. 21431, 2015 revised 2017). NER.

Caswell, D. (2005). *Handlemuligheder i socialt arbejde; et casestudie om kommunal frontlinjepraksis på beskæftigelsesområdet [Options in action in social work – A case study on municipal front line practice within the field of employment]* [Doctoral dissertation]. AKF-forlaget.

Christensen, A. M. (2018). *Grønlandsk økonomi. Stærk vækst og mangel på arbejdskraft [Greenland economy. Shortage of labour and strong growth]*. Danmarks Nationalbank.

Dahl, K. M., Skov Kloppenborg, V. H., & Mau Pedersen, N. J. (2020). *Kortlægning af døgninstitutionsområdet i Grønland [Mapping of day institutions in Greenland]*. VIVE.

Department of Finance. (2019). *Politisk-Økonomisk Beretning 2020 [Political-Economic report]*. FM 202/10.

Freud, D. (2007). *Reducing dependence, increasing opportunity: options for the future of welfare to work*. Department for Work and Pensions.

Grønlands Økonomiske Råd. (2020). *Grønlands økonomi, foråret 2020 [Greenland economy. Spring 2020]*. Grønlands Økonomiske Råd.

Grønlands Selvstyre. (2019). *Forslag til finanslov 2020* [Budget proposal 2020]. Grønlands Selvstyre.

Grønlands Selvstyre. (2020). *Forslag til finanslov 2021* [Budget proposal 2021]. Grønlands Selvstyre.

Holt, H., Thuesen, F., & Casier, F. (2019). *Evaluering af Majoriaq-centrenes arbejdsmarkedsindsats [Evaluating the efforts of job centres in Greenland]*. VIVE.

Kim, J., Lotz, J., & Mau Pedersen, N. J. (eds.) (2013). *Balance between decentralization and merit. The Copenhagen Workshop 2011*. Korea Institute of Public Finance and Danish Ministry of Economic Affairs and the Interior.

Koch, A., & Rørbech, M. (1985). *Kontanthjælpsforsøget – Frigjorte ressourcer og forbedret social arbejde? [The social security experiment - Released resources and improved social effort]*. AKF & SFI – det Nationale Forskningscenter for Velfærd.

Lindsay, C., & Mailand, M. (2004). Different routes, common directions? Activation policies in Denmark and the UK. *International Journal of Social Welfare, 13*(7), 195–207.

Lovgivning.gl. (2006). *Landstingsforordning nr. 15 af 20. November 2006 om offentlig hjælp*. http://lovgivning.gl/lov?rid=[BE42033F-E5BB-4199-8E76-91F7C9401BBF]

Lovgivning.gl. (2019). *Selvstyrets bekendtgørelse nr. 12 af 9. Oktober 2019 om skatteudligning og fælleskommunal sat i 2020*. http://lovgivning.gl/Services/ GeneratePdf?pdfurl=%2Fda-DK%2FLov%3Frid%3D%257b4213FD69-7F1C-4941-91F2-4F3C9165178B%257d&attachment=true&attachmentname=Selv styrets%20bekendtg%C3%B8relse%20nr.%2012%20af%209.%20oktober%20 2019%20om%20skatteudligning%20og%20f%C3%A6lleskommunal%20skat%20 i%202020

Lovgivning.gl. (2020). *Selvstyrets bekendtgørelse af 21. Februar 2020 af Inatsisartutlov om førtidspension*. http://lovgivning.gl/lov?rid=%7B43DAE349-01EE-4135-AAD1-0EB03DF08E19%7D

Mau Pedersen, N. J. (2017). *Et hensigtsmæssige refusionssystem – hvilke hensyn må tages? Rapport-notat til projekt om Analyse af offentlig hjælp* [An appropriate reimbursement system – which considerations to take into account?]. VIVE.

Mau Pedersen, N. J., Christensen, N., Houlberg, K., Ejersbo, N., & Kolodziejzcyk, C. (2019a). *Refusionsomlægning på beskæftigelsesområdet. Økonomiske og administrative konsekvenser* [Reimbursement reform for labour market schemes. Economic and administrative consequences]. VIVE.

Mau Pedersen, N. J., Schøler Kollin, M., Line Tenney Jordan, A. L., & Hønge Flarup, L. (2016). *Inkorporering af FN's børnekonvention i Grønland* [Incorporating UN Children's Convention in Greenland]. KORA.

Mau Pedersen, N. J., Seier Petersen, J., & Højgaard Lindeberg, N. (2019b). *Analyse af offentlig hjælp [Analysis of public assistance system]*. VIVE.

Naalakkersuisut. (2016). *Notat om overvejelser til reform omkring offentlig hjælp [Note on considerations about a reform of public assistance system]*. Naalakkersuisut.

Naalakkersuisut. (2020a). *Koalitionsaftale. Vækst, stabilitet og tryghed. Mellem Siumut, Demokraatit og Nunatta Qitornai* [Agreement coalition. Growth, stability and security. Agreement between Siumut, Demokraatit og Nunatta Qitornai]. Politisk Program.

Naalakkersuisut. (2020b). *Budgetsamarbejdsaftale mellem selvstyret og kommunerne* [Budget-agreement between the Central Government and the municipalities]. Politisk Program.

Skatte- og Velfærdskommissionen. (2011). *Vores velstand og velfærd – kræver handling nu. Skatte- og Velfærdskommissionens betænkning* [Our welfare society demands action now. Tax and Welfare Commission report]. Naalakkersuisut.

Socialreformkommissionen. (1997). *Betænkning 1997. 15. April 1997.*

STAR. (2014). *Empowerment i det beskæftigelsesrettede arbejde* [Empowerment in the employment efforts]. Styrelsen for Arbejdsmarked og Rekruttering.

7 Individual factors associated with labour market participation in Greenland

Rasmus Lind Ravn and Laust Høgedahl

Abstract

In this chapter, we explore individual level factors that are associated with employment outcomes in Greenland using data from the Greenlandic Labour Force Survey. Research findings related to such factors are well known in international research but much more limited in a Greenlandic context. From a policymaker perspective, information about such factors is crucial in designing and targeting policies that can improve employment outcomes. Through logistics regression analyses, we find that educational attainment is strongly associated with employment outcomes. Geography and participation in the educational system also affect employment outcomes in Greenland. Our results indicate a need for increasing the educational attainment of the Greenlandic population, especially considering the educational attainment is quite low in an international perspective.

7.1 Challenges in relation to employment in Greenland

The chapter investigates labour market participation in Greenland based on unique data from the first Greenlandic labour force survey (G-AKU). The data are comparable with international LFS statistics, making cross-national comparisons of Greenlandic labour market conditions possible. We focus on individual factors for labour market participation and not as much on the institutional set-up (for these, see Chapters 4 and 6).

Our motivation for looking at individual factors associated with labour market participation is twofold. First, Greenland has seen an economic upturn in recent years, creating a labour shortage in the country despite a rather large part of the working population being outside of the labour force (see Chapter 2). Second, we can argue that the Greenlandic labour market is unique in terms of structural and cultural features, making an international comparison of labour market participation interesting and relevant. The existing literature dealing with labour market participation in Greenland is mainly based on qualitative and, especially, ethnographic data. By applying a quantitative approach, we are able to supplement and test the existing findings on labour market participation in Greenland.

The aim of this chapter is, therefore, to investigate individual factors associated with labour market participation in Greenland by utilising quantitative survey data based on descriptive and model-based statistics. In spite of the growing interest in the Arctic area, and Greenland in particular, the body of existing literature on key labour market features, such as employment and labour supply is scarce, mainly because of the small population of only 55,000 inhabitants. However, relevant data has also been an issue – albeit with limited resources – Statistic Greenland has developed a comprehensive register data collection including labour market statistics in line with ILO standards. Yet, Greenland is one of the few countries affiliated with the European Union that are not part of the Labour Force Survey (LFS). However, in 2019, the Greenlandic Government funded a Greenlandic version of the LFS (see Chapter 5 for more on the survey). The aim of the survey is to provide supplementary data on the Greenlandic labour supply in addition to the register data provided by Statistic Greenland. More information on the Greenlandic supply is urgently needed in a time of economic development and labour shortage. Since 2019, three large airport projects have been underway simultaneously, making the demand for labour even more pressing in the country (The Economic Council of Greenland, 2019; The National Bank of Denmark, 2019). Greenland has – despite a booming economy and growing employment during the past years – a relatively large part of the population who, for different reasons, are not in employment (see Chapter 5).

Therefore, the outline of the chapter is as follows. In the next section, the distinct structural, institutional, and not least, cultural features of the Greenlandic labour market are presented by consulting the existing literature. We focus on what is known about individual factors associated with labour market participation. We then turn to theories on barriers for labour market participation, with a point of departure in international research related to studies using LFS data, followed by a brief introduction to our data and method. Later, we turn to our research results, and in the final section, we conclude that despite having a rather unique labour market, the same known barriers for labour market participation are found in Greenland compared to European countries. However, we also find some interesting peculiarities.

7.2 The Greenlandic labour market – a context description

When addressing the literature regarding individual factors associated with labour market participation in Greenland, a number of studies can be identified. First, the geography and the size of the country make mobility and transportation challenging. It is often said that Greenland does not have one labour market but many isolated markets for labour due to the vast distances and logistical challenges (Olsen, 2005). When surveying individual factors connected to labour market participation, geography is an important variable. Seasonal differences have diverse effects in different parts of the country. Most of the tertiary economy, such as private service, is based in Nuuk, making this local labour market less affected by weather and seasonal changes. Greenland's colonial history with Denmark is also important to bear in mind when looking at individual factors for labour market participation. The political economy of the country has transformed from a provision-based economy focused on fishing and hunting over colonial trade to a mixed economy today (see Chapter 3). Hence, Greenland has a rather large public sector, making up approximately 36% of the employment (Statistics Greenland, 2019). In addition, many of the large companies are owned by the state, including Royal Greenland (fishing) and Air Greenland (air logistics). The private economy is heavily dependent on fishing and its downstream industries, making the Greenlandic economy vulnerable to fluctuation in fish prices and stocks. Therefore, new areas of business are being pursued, such as mining, adventure tourism, renewable energy, and pure water export, combined with labour market reforms aimed at increasing the labour supply.

The historical and structural features of the Greenlandic labour market are pivotal for understanding the Greenlandic labour force. Poppel (2005) shows how the Inuit culture of male hunters, sealers, whalers, and fishermen is still embedded in the modern labour make today. Many male workers consider factory work in the fishing industry as female work that does not resonate with the perception of a male breadwinner. Most self-employed in Greenland are males fishing from small boats, but the share of self-employed women is much lower (Høgedahl & Krogh, 2019). Different cultural traditions related to gender are important factors for the Greenlandic labour market (see Chapter 8). The long Inuit tradition as hunters and gatherers based on a provision-based economy are also affecting the general notion of "work". The term "work" is pretty well understood as an action in which a worker sells his or her labour to an employer for a given price to obtain a certain living standard. When the notion of "work" differs from the general understanding, "unemployed" becomes equally different, as Hansen and Tejsner (2016) show from a study in North West Greenland.

7.3 Factors associated with employment in Greenland and internationally – a brief review of the literature and empirical patterns

A great deal of research related to which factors are associated with employment has been carried out internationally. In this section, we outline the findings of these studies, as well as studies of Greenland, in order for us to generate hypotheses about which factors are associated with employment in

Greenland. Our focus is primarily on eliciting existing empirical patterns of employment and unemployment. We, therefore, focus on existing empirical research about Greenland, supplemented with international studies when needed. We do this to determine, in the conclusion, whether Greenland is an exception from the international trends or if the country is similar to the majority of countries in terms of which factors influence employment outcomes. We structure the review according to gender, educational attainment, age, geography, and participation in education, which are the factors we examine as part of our analysis.

Statistics Greenland, the Government of Greenland, and the Economic Council of Greenland regularly publish reports and statistics that are relevant when investigating factors associated with employment. We use these primary sources for outlining the Greenlandic research on the subject.

7.3.1 Gender and employment

Internationally, women generally have a lower employment rate than men, the so-called employment gap (OECD, 2012). Gender employment gaps are, however, rather small in the Nordic countries, compared to the majority of OECD countries. This is due to the Nordic welfare state model providing childcare and eldercare, making a higher employment rate for women possible. In addition, when employed, women also tend to have lower earnings, compared to their male counterparts (OECD, 2017). The employment rate of Greenlandic women is also lower than that of the Greenlandic men. However, differences are not extremely pronounced, with a difference of roughly 8 percentage points (Høgedahl & Krogh, 2019, p. 25). However, the gender differences in employment rates might disappear when taking account of, for instance, educational attainment, which we will investigate in the following analysis. In fact, evidence suggests that the gender employment gaps narrow (or even vanish) with increasing educational attainment (OECD, 2012). This is connected to the next theme in our review, which is precisely educational attainment.

7.3.2 Educational attainment and employment

In the vast majority of OECD countries, unskilled workers, or groups with low educational attainment, are underrepresented in employment statistics (OECD, 2015). As the educational level increases, so does the share who are in employment. As such, completing an education is a way of improving your chances of being employed. This is most definitely the case in Greenland because unemployment among people with further education is almost non-existent in Greenland (The Economic Council of Greenland, 2017). Therefore, we expect that educational attainment will be positively associated with employment status, even while taking account of other factors, such as gender or age.

7.3.3 Age and employment

The Greenlandic Government has a strong focus on reducing youth unemployment, as young Greenlanders are overrepresented in official unemployment statistics (Greenlandic Ministry of Finance, 2019; Government of Greenland, 2015). In relation to this, existing international research finds that age is related to employment outcomes. Young people have a lower probability of employment than prime-age workers for a number of reasons do. First, they are the primary target group of the educational system, which naturally lowers the share in employment. Furthermore, upon graduation from an education, young people will have formal qualifications, but they often lack work experience when competing for jobs with prime-age workers. Consequently, youth employment will be lower than that of prime-age workers, which we also expect will be the case in Greenland.

Likewise, the employment rate of older workers is typically lower than that of prime-age adults. In the EU28 countries, the employment rate among 55- to 64-year-olds in the second quarter of 2019 (the time at which the Greenlandic LFS was conducted) was 10 percentage points lower than that of the entire working age population (Eurostat [lfsq_ergan]). A plausible explanation for this is that, if older workers experience unemployment, they use it as a way of early retirement, thereby exiting the labour force (Choi et al., 2015). Furthermore, Danish research shows that the unemployment rate of seniors is higher than the general unemployment rate (Qvist & Jensen, 2020). A plausible explanation is that employers are reluctant to hire older workers if there are young equally qualified candidates for the job (SFI, 2016). Senior unemployment is particularly pronounced among unskilled and skilled workers and other groups with low educational attainment (Qvist & Jensen, 2020). This should be particularly relevant in Greenland because the educational level of the working age population in Greenland is quite low in comparison with other countries (cf. Chapters 2, 4, and 9). Therefore, we expect a lower unemployment rate among the older working age population.

7.3.4 Geographic patterns of employment and unemployment

Geography plays a key role in relation to employment and unemployment. Opportunities for employment naturally varies across geographic areas. In Greenland, the geographic mobility of the population is extremely high – even compared to other countries (Mobilitetsstyregruppen, 2010). The Greenlandic Economic Council, however, finds that the Greenlandic mobility patterns are not from high unemployment areas to low unemployment areas, which would indicate that people do not move for job opportunities (The Economic Council of Greenland, 2018, p. 41).

Data from Eurostat shows that the employment rate in rural areas among EU28 are slightly lower than in areas with higher degrees of urbanisation. Access to jobs and job opportunities are generally scarcer in rural areas

(Détang-Dessendre & Gaigné, 2009). Furthermore, the unemployed living in remote areas are often less active in searching for a job because the costs of job searching are greater for these people (Kingdon & Knight, 2006). Furthermore, the employability, in terms of educational attainment and qualifications, of people living in remote areas is often lower than that of the general population, and they often lack access to employment services, hampering their job chances (Lindsay et al., 2003). Based on the above, we expect to find a lower employment rate among Greenlanders living in other places than Nuuk and Sisimiut (the largest and second largest town/city in Greenland).

7.3.5 Participation in education and employment

Generally, the employment rates of students are lower than that of the general working age population. In 2016, 54% of young people in the EU28 countries did not work at all while studying (Eurostat, 2018). This makes sense, as students spend their time studying and not working.

Therefore, we expect to observe a low share of students in employment in Greenland. Furthermore, we hypothesize this to be particularly pronounced in Greenland because Greenland offers a relatively generous monthly student grant (income transferring benefit for participating in education) for people enrolled in an education. In fact, the Greenlandic student grant is quite high, compared to other Nordic countries. In theory, this should make part-time work less of a necessity than in the majority of countries. Therefore, we expect that being a student will lower the likelihood of being employed.

7.4 Data and methods

Our data for the analysis is based on the first standardised LFS conducted in Greenland. For more information regarding the survey, please refer to Chapter 5 in this book.

To investigate which factors are associated with employment and barriers for labour market participation in Greenland, we run three binary logistic regression models predicting odds ratios for the covariates. An odds ratio larger than "1" indicates a higher likelihood of employment compared to the reference category, whereas an odds ratio lower than "1" indicates a lower likelihood of employment.

The covariates we include are all chosen to reflect our hypotheses developed above. The covariates are chosen because existing research shows that age, gender, educational attainment, and geography are valid predictors of employment status. To investigate if this is also the case in Greenland, we include the following covariates in the final regression model.

- Age
- Gender

- Educational attainment
- Settlement or town/city
- Nuuk/Sisimiut or other
- Participation in education

The classification of "settlement or town/city" requires some additional explanation. In Greenland, a large share of the population lives in so-called "settlements" which are populated areas that are not large enough to be classified as a town or a village. A settlement can be located just outside a town or it can be located in a rather isolated area. Our variable of interest thus measures whether the respondents live in a settlement, town, or city (The only city in Greenland per definition is Nuuk because it is the only place that has a cathedral). Of the respondents in the survey, 12.9% live in a settlement and 87.1% live in a town or a city.

We have also included a variable that measures whether the respondents live in one of the two largest towns or cities. The largest city is Nuuk, with roughly 18,300 inhabitants, and the second largest town is Sisimiut, with roughly 6,000 inhabitants.

Odds ratios and coefficients in logistic regression models can, however, be difficult to interpret because they do not directly relate to the actual probability of achieving the outcome (Ravn, 2019) – in our case, being employed. Based on our final regression model, we therefore predict the marginal probabilities (in percentages) of being in employment, while controlling for the covariates included in our final model. For instance, this enables us to estimate the likelihood of employment for men and women, while taking account of age, educational attainment, and geography.

To test for multicollinearity, we conducted a variance inflation factor test (VIF-test) by running a linear regression model including the same dependent variable and the same independent variables. The results of the test showed no signs of multicollinearity (mean VIF value = 1.16, largest VIF value = 1.36).

7.5 Individual factors associated with labour market participation in Greenland – empirical results

As suggested above, we expect that gender, educational attainment, geography/place of residence, and participation in education will affect employment outcomes in Greenland, as this is the case in the vast majority of OECD countries. Therefore, the analysis begins by presenting descriptive statistics on the dependent variable and the aforementioned covariates. These results can be seen in Table 7.1.

From Table 7.1, we see that 78.3% of the respondents were employed in the reference week. Employment includes wage earner employment and self-employment. The Greenlandic employment rate is thus quite similar to that of Denmark and other European countries (Høgedahl & Krogh, 2019).

Table 7.1 Descriptive statistics of the dependent variable and the covariates

Employed	78.3%
Gender	
Men	51.4%
Women	48.6%
Educational attainment	
Unskilled (at most lower secondary school)	44%
Further education	9.3%
Skilled worker/Vocational education	30.2%
No of the above	16.5%
Settlement or town/city	
Settlement	12.9%
Town or city	87.1%
Place of residence	
Nuuk/Sisimiut	40.9%
Other town or settlement	59.1%
Participation in education	
Currently in education	7.3 %
Not currently in education	92.8. %

Of our respondents, 51.4% were men and 48.6% were women. The Greenlandic population consists of 53.1% men and 46.9% women. Focusing on the educational attainment of the respondents, we see that a very large share (44%) have only compulsory schooling or a general upper secondary school degree. Compared to, for instance, the Nordic countries, we see a very large share with limited educational attainment (cf. Figure 6.4 in Chapter 6). Of the respondents, 9.3% have completed further education, for instance a bachelor's or a master's degree, and 30.2% of the respondents are skilled workers who completed a vocational education. Lastly, 16.5% have not completed any of the educations listed above, which in many cases should be interpreted as no educational attainment.

Turning to geography, we find that 40.9% of our respondents live in either Nuuk or Sisimiut, whereas the rest (59.1%) live in a different town or settlement.

Among the participants of the survey, 7.25% are currently enrolled in an education. The majority of students are between the ages of 17 and 29 years.

We also summarise the share of the respondents participating in education. From Table 7.1, we see that 7.3% of the respondents were enrolled in an education. Finally, we find the mean age of our respondents was 39.4 years (not shown in Table 7.1).

Having outlined the descriptive presentation of the data, we investigate the binary associations between our covariates and the dependent variable of employment. For presentational purposes, we have recoded the age variable into categories, but in the regression analyses, we use it in its original continuous form.

As can be seen from Table 7.2, there are some differences in employment between the different groups. For instance, we see that only 53.1% of the

Table 7.2 Binary associations between covariates and employment

	Percentage in employment
Age	
17–20	53.1
20–29	73.4
30–39	78.5
40–49	85.6
50–59	85.7
60–64	72.6
Gender	
Men	82.4
Women	73.9
Educational attainment	
Unskilled (only compulsory school or upper secondary school)	66.8
Further education	88.0
Skilled worker/Vocational education	91.1
None of the above	79.7
Settlement or town/city	
Settlement	74.1
Town or city	78.9
Nuuk/Sisimiut or other place of residence	
Nuuk/Sisimiut	80.9
Other town or settlement	76.4
Participation in education	
Currently enrolled in education	15.4
Not enrolled in education	83.2
Total	78.3
N	897

respondents below 20 years old are in employment. The obvious explanation is that they are still in school or undertaking education. Employment rates rise through the 20s and 30s and tops in the age groups between 40–49 years and 50–59 years. Then, it sharply drops for the 60- to 64-year-olds. This is expected because employment rates of seniors (60–64 years) are generally lower than those for the remaining working age population in most European countries (Eurostat [lfsa_ergan]). Interestingly, the Greenlandic LFS employment rate of 2019 for the 60- to 64-year-olds (72.6%) was much higher than the EU28 average for the same age group in 2019 (46.0%).

In relation to gender differences, we see that the employment rate of men is almost 9 percentage points higher than that of women. This employment gap of 9 percentage points is not unusually high. In fact, an employment gap of 10 percentage points between men and women is common in many European countries. In 2019, the EU28 employment gap between men and women amounted to 10.3 percentage points – larger than the Greenlandic employment gap. In addition, the female Greenlandic employment rate is higher than many EU countries, and much higher than the EU28 countries (73.9% in Greenlandic and 64.1% in EU28).

Internationally, unskilled workers and people with low educational attainment have lower employment rates than their counterparts with further education (OECD, 2019). This pattern is also evident in Greenland. Concerning educational attainment, we see that those with only compulsory schooling or a general upper secondary school degree have a lower than average employment rate of 66.8%, which is almost 12 percentage points lower than the average of 78.3%. The groups with further education or vocational education have quite high employment rates (88.0% and 91.1%, respectively). The same is also the case for the group who does not fall into any of the listed educational categories (79.7%).

Turning to the differences in employment between settlements on the one hand and towns/cities on the other, we surprisingly find rather small differences. The employment rate is roughly 4 percentage points higher in towns and cities compared to settlements.

Finally, we inspect the differences in employment between the two largest towns/cities (Nuuk and Sisimiut) and other places of residence. Here, we also find a minor difference of approximately 4 percentage points, with employment being the highest in Nuuk and Sisimiut, which could potentially reflect that demand for labour is strong in these places.

To further investigate if these associations hold when taking account of other factors, we next run three logistic regression models with employment status as the dependent variable (Table 7.3).

In Model 1, we include only two covariates: age and gender. Both variables become highly significant. In Model 1, a higher age is associated with a higher likelihood of being in employment. As mentioned earlier, this is likely because young people are still in education instead of being in employment. Gender also becomes highly significant in Model 1. Women have a lower likelihood of being in employment compared to men (odds ratio = 0.6).

In Model 2, we add educational attainment. Here, it is worth noting that age no longer is significant. When controlling for educational attainment, the association between age and employment status disappears, indicating that the lower employment rate of young people is because they have not yet finished their education. The gender differences, however, persist. All categories of educational attainment have odds ratios above "1", and all categories are significant at a 0.05 level, indicating that labour market outcomes are better for all groups with some education compared to the reference category (the group with only compulsory school or general upper secondary school). However, it is worth noting that Greenlandic women still have a lower likelihood of employment, even when controlling for educational attainment.

In the third model, we add geography in the form of our variables measuring if the respondents live either in a city/town or in a settlement and if they live in either Nuuk or Sisimiut or in another place of residence.

Table 7.3 Logistic regression with employment status as the dependent variable (odds ratios)

	Model 1 (Odds ratios)	Model 2 (Odds ratios)	Model 3 (Odds ratios)	Model 4 (Odds ratios)
Age	1.025*** (0.006)	1.009 (0.007)	1.010 (0.007)	0.995 (0.007)
Gender (ref. male)	0.614** (0.102)	0.592** (0.101)	(0.584)** (0.100)	0.716 (0.134)
Education				
Unskilled		Ref.	Ref.	Ref.
Further education		3.676*** (1.325)	3.467*** (1.257)	3.181** (1.275)
Vocational education		4.617*** (1.154)	4.550*** (1.139)	3.838*** (1.009)
None of the above		1.734* (0.416)	1.748* (0.420)	1.404 (0.353)
City/town or settlement (ref. city/town)			0.832 (0.204)	0.720 (0.182)
Nuuk/Sisimiut or other town or settlement (ref. other town or settlement)			1.295 (0.234)	1.677** (0.344)
Participating in education (ref. not participating)				0.040*** (0.015)
Constant	1.838	1.984	1.773	3.896
Pseudo R²	0.028	0.082	0.085	0.196
N	897	897	897	897

Note: Standard errors in parentheses.

$*p \leq 0.05$, $**p \leq 0.01$, $***p \leq 0.001$.

Surprisingly, neither of the two variables becomes significant. This is also the case when including only one of the variables in the regression (not shown). Theoretically, employment opportunities should be best in the more densely populated areas. In addition, most educational institutions are situated in Nuuk and Sisimiut, which may affect the employment rate negatively, especially as young people might be in education. However, Model 3 qualitatively shows the same as Model 2. Women are still less likely to be in employment than men, and having had further schooling than what is mandatory in Greenland increases the likelihood of employment. Model 3 shows no evidence of geography influencing the likelihood of employment.

This, however, changes in Model 4, where we add a variable measuring whether the respondents are currently participating in education.

Controlling for participation in education is particularly relevant when young people constitute a significant part of your sample, as the primary target group for education is young people. In addition, Chapter 9 stresses that a larger share of Greenlandic women are studying compared to the men. The lower employment rate of women could simply reflect that they are enrolled in the educational system to a larger extent than men are. Lastly, the seemingly equal likelihood of employment in relation to geography or place of residency might be spurious because education opportunities are limited outside of the larger towns. In other words, not having controlled for participation in education might shroud the fact that employment opportunities are, in reality, better in certain parts of Greenland, namely the larger towns.

Inspecting the results in Model 4, they seem to support these notions. Women do not have a lower likelihood of employment than their counterparts do when taking account of geography and participation in education. Thus, it would seem that the above suspicion is correct. Women in Greenland have a lower employment rate than men because they participate in education to a further extent than men do.

We furthermore find that participating in education lowers the likelihood of employment. As elaborated upon in Chapter 9, this makes sense because students typically are enrolled in full-time education and receive a monthly student grant (an income transferring benefit for participating in education).

An interesting change happens from Model 3 to Model 4. Geography, more precisely the variable measuring if the respondent lives in Nuuk or Sisimiut, becomes statistically significant when controlling for current participation in education. Thus, the likelihood of employment is actually higher for residents in Nuuk or Sisimiut than the rest of the country. This association between living in the two largest populated areas of Greenland and employment was suppressed by the fact that a higher share of young people in Nuuk and Sisimiut are enrolled in education. Considering all the variables in the model, the probability of employment is higher in Nuuk and Sisimiut than the rest of the country.

As mentioned in the methods section, odds ratios are difficult to interpret because they do not directly relate to the likelihood of the outcome – they relate to the likelihood of the outcome compared to the reference category. To ease and simplify interpretation, we predict the marginal probabilities of being in employment based on the fourth regression model, as shown in the Table 7.4.

When interpreting Table 7.4, it is crucial to bear in mind that the results are based on the fourth regression model. This implies that we, for instance, estimate the likelihood of male employment while controlling for the other covariates in Model 4.

Focusing on the gender differences, we see that men have a marginal probability of employment of 80%, whereas women have a marginal probability of employment of 76%. The confidence intervals overlap, indicating

Table 7.4 Post estimation based on the fourth logistic regression model

	Marginal probability	Delta-method std. error	Z-value	p-value	95% Confidence interval Lower bound	Upper bound
			Sex			
Men	0.80	0.02	49.09	0.000	0.77	0.84
Women	0.76	0.02	41.16	0.000	0.72	0.80
			Age			
20	0.80	0.02	39.56	0.000	0.76	0.84
30	0.79	0.01	55.61	0.000	0.76	0.82
40	0.78	0.01	64.43	0.000	0.76	0.81
50	0.78	0.02	47.24	0.000	0.74	0.81
60	0.77	0.02	31.48	0.000	0.72	0.82
		Settlement or city/town				
Settlement	0.70	0.013	60.77	0.000	0.76	0.82
City/town	0.74	0.035	21.53	0.000	0.68	0.81
		Nuuk/Sisimiut or other town or settlement				
Other place of residency	0.76	0.02	46.08	0.000	0.73	0.79
Nuuk/Sisimiut	0.82	0.02	45.69	0.000	0.79	0.86
Highest completed education						
Unskilled (primary and secondary school)	0.71	0.02	31.43	0.000	0.67	0.75
Further education	0.86	0.04	24.32	0.000	0.79	0.93
Vocational education	0.88	0.02	45.51	0.000	0.84	0.92
No of the above	0.76	0.03	23.98	0.000	0.70	0.82
		Participation in education				
Not enrolled education	0.83	0.01	64.56	0.000	0.80	0.85
Currently enrolled in education	0.21	0.01	3.64	0.000	0.10	0.32

no statistically significant differences. It would thus seem that women are *not* more disadvantaged on the Greenlandic labour market than their male counterparts are.

Regarding age, the marginal probabilities of employment have been estimated at 10-year points. The age variable has still been used in its original continuous form. We see that the estimated likelihood of employment differs slightly depending on age. However, all confidence intervals overlap, signifying that there are no real differences depending on age. Neither young people nor seniors are less likely to be in employment than prime-age workers are when taking account of the other variables.

The same is the case regarding our first variable: geography. We find no differences between living in a settlement or a town/city. However, we do find that employment chances are better in Nuuk and Sisimiut compared to other parts of Greenland – bear in mind that we do control for education levels. As already mentioned, in relation to Table 7.3, the likelihood of employment is higher if the respondent lives in Nuuk or Sisimiut, which plausibly reflects that demand for labour is higher in these places.

In relation to educational attainment, we see that unskilled workers have a 71% likelihood of being in employment, whereas those with further education or a vocational education have a likelihood of 86% and 88%, respectively. From the confidence intervals regarding education, we see that unskilled workers have a significantly lower likelihood of employment than people with further education or a vocational education do. However, we cannot conclude that unskilled workers have a lower likelihood of employment than the group not falling into any of the educational categories. Furthermore, we find no differences in the likelihood of employment between the last three educational categories.

Finally, what matters most for your chance of being in employment in Greenland, in absolute terms, is whether you are currently in education. Those who are currently in education have only a 21% chance of being in employment, whereas those who are not participating in education have a 83% probability. This finding is not surprising because education for the most part is full-time, leaving little time for work.

7.6 Conclusions – individual factor associated with employment in Greenland

In this chapter, we have investigated which factors are associated with a higher probability of employment in Greenland. In other words, we explore potential barriers for employment on the Greenlandic labour market. Prior to our study, no LFS had been conducted in Greenland. The Greenlandic labour market is thus a vastly underexplored research area, and knowledge about individual factors related to employment is limited, despite growing demands for labour. Our study thereby contributes to the existing research mostly based on qualitative data by investigating which factors are empirically associated with a higher chance of employment in Greenland compared to other countries.

In our descriptive analyses, we find that that young Greenlanders (17–29 years) and older Greenlanders (60–64 years) have lower employment rates than prime-age workers (30–59 years). This is also a pattern found in most other countries. In addition, Greenlandic women have a lower employment rate than their male counterparts do – an empirical pattern also found in almost every OECD country. Similarly, unskilled Greenlandic workers are in employment to a much lesser extent than the Greenlanders who have completed either further education or a vocational education. The share in employment thus increases with educational attainment. In this regard, Greenlandic is also strikingly similar to most other countries. We find some minor geographic differences in employment rates. Employment rates are higher in towns/cities than in settlements, which resonates with other research findings. The same was the case for respondents living in Nuuk (the capital) or Sisimiut (the second largest town in Greenland) compared with living in the rest of the country. Lastly, and in line with existing international research, we find a much lower share of Greenlandic students in employment compared to the remaining working age population. The above are merely descriptions of binary relationships. However, these relationships or associations might vanish when we consider other factors. The second part of our results explored these multivariate relationships through logistic regression analyses.

In the first three regression models (controlling for age, gender, and educational attainment), we found no evidence that geography influenced the likelihood of being in employment. To some degree, this was an unexpected finding, when taking into consideration that existing research consistently finds that job opportunities are fewer in rural areas. When controlling for participation in education, geography, however, turned out to play a role in employment. The likelihood of employment is in fact larger in Nuuk and Sisimiut. However, this was shrouded by the fact that these two cities/towns have a larger share of students, which suppressed the fact that employment opportunities are better in these areas. Intuitively, this makes sense, as the demand for labour is often stronger in urban areas compared to rural areas.

Initially, we also found a lower employment rate of women (compared to men) – even when taking account of educational attainment, age, and geography. Gender differences, however, evened out when controlling for participation in education. The lower employment rate of women is thus due to the fact that Greenlandic women are participating in education to a larger extent than Greenlandic men are. As such, Greenlandic women are not more disadvantaged in relation to employment opportunities than their male counterparts are.

As mentioned earlier, the Government of Greenland has a strong focus on increasing youth employment. Our results, however, show that neither young people nor seniors are less likely to be in employment when considering the other factors in the fourth regression model. Young people are less likely to be in employment because a larger share of them are in education compared to the rest of the working age population, which leads us to educational attainment. In relation to this, we find that educational attainment does, indeed, influence the likelihood of being in employment. Furthermore, having any type of education increases the likelihood of being in employment compared to having only compulsory school or a general upper secondary school degree.

Our results have a series of possible implications. In relation to education, the educational level of the Greenlandic population is quite low in an international perspective (see also the discussion in Chapter 2). The strong association between educational attainment and employment stresses the need for increasing the educational level of the Greenlandic population.

At first glance, the Greenlandic labour market seems unique, with a very large share of public sector employees, a private sector driven primarily by fishing and downstream industries, and a labour market that cannot be conceived of as a single labour market, but many different labour markets due to the vast geographic distances in the country. As such, the Greenlandic labour market is unique. However, concerning factors that are associated with employment, we find that Greenland is, indeed, quite similar to most other countries.

Our study has contributed to the limited knowledge about the Greenlandic labour market and factors associated with employment. However, more research into demand for labour in the Arctic and the causal mechanisms driving the differences in employment in Greenland is much needed.

References

Choi, S., Janiak, S., & Villena-Roldán, B. (2015). Unemployment, participation and worker flows over the life-cycle. *The Economic Journal, 125*(589), 1705–1733.

Détang-Dessendre, C., & Gaigné, C. (2009). Unemployment duration, city size, and the tightness of the labor market, *Regional Science and Urban Economics, 39*, 266–276.

Eurostat. (2018, January). *Young people on the labour market statistics*. Eurostat. https://ec.europa.eu/eurostat/statistics-explained/index.php/Young_people_on_the_labour_market_-_statistics

Government of Greenland. (2015). *A safe labour market. Employment strategy 2015*. https://naalakkersuisut.gl/~/media/Nanoq/Files/Publications/Arbejdsmarked/DK/Beskaeftigelsesstrategi_DK_01102015.pdf

Greenlandic Ministry of Finance. (2019). *Political economic report, Ministry of Finance*. https://naalakkersuisut.gl//~//media/Nanoq/Files/Attached%20Files/Finans/DK/Politisk%20Oekonomisk%20Beretning/P%C3%98B2019%20DK.pdf

Hansen, A. M., & Tejsner, P. (2016). Challenges and opportunities for residents in the Upernavik District while oil companies are making a first entrance in Baffin Bay. *Arctic Anthropology, 53*(1), 84–94.

Høgedahl, L., & Krogh, C. (2019). *Den Grønlandske AKU-undersøgelse (G-AKU)* [The Greenlandic Labour Force Survey]. *Forskningsrapport*. Aalborg Universitet.

Kingdon, G., & Knight, J. (2006). The measurement of unemployment when unemployment is high. *Labour Economics, 13*, 291–315.

Lindsay, C., McCracken, M., & Mcquaid, R. W. (2003). Unemployment duration and employability in remote rural labour markets. *Journal of Rural Studies, 19*, 187–200.

Mobilitetsstyregruppen (2010). *Mobilitet I Grønland. Sammenfatning af hovedpunkter fra analysen af mobilitet i Grønland [Mobility In Greenland. Summary of main points from the analysis of mobility in Greenland]*. Mobilitetsstyregruppen.

OECD (2012). *Closing the gender gap: Act now*. OECD Publishing. http://dx.doi.org/10.1787/9789264179370-en

OECD (2015). OECD Employment Outlook 2015. OECD Publishing. http://dx.doi.org/10.1787/empl_outlook-2015-en

OECD. (2017). OECD Employment Outlook 2017. OECD Publishing. http://dx.doi.org/10.1787/empl_outlook-2017-en

OECD. (2019). Education at a glance 2019: OECD indicators. OECD Publishing. https://doi.org/10.1787/f8d7880d-en

Olsen, J. L. (2005). Udfordringer til det grønlandske arbejdsmarked [Challenges to the Greenlandic labor market]. In A. V. Carlsen (Ed.), *Arbejdsmarkedet i Grønland – fortid, nutid og fremtid*. Nuuk: Ilisimatusarfik.

Poppel, M. (2005). Barrierer for grønlandske mænd på arbejdsmarkedet [Barriers for Greenlandic men in the labor market]. In A. V. Carlsen (Ed.), *Arbejdsmarkedet i Grønland – fortid, nutid og fremtid* [The labour market in Greenland - past, present and future, 2005] (pp. 125–140). Nuuk: Ilisimatusarfik.

Qvist, J. Y., & Jensen, P. H. (2020). Arbejdsløsheds- og beskæftigelsesmønstre blandt seniorer [Unemployment and employment patterns among seniors]. In P. H. Jensen (Ed.), *Seniorarbejdsliv*. Frydenlund Academic.

Ravn, R. (2019). Testing mechanisms in large-N realistic evaluations. *Evaluation, 25*(2), 171–188. https://doi.org/10.1177/1356389019829164

SFI. (2016). *Evidens om effekten af indsatser for ledige seniorer – en litteraturoversigt* [Evidence about the effect of initiatives for unemployed seniors - a literature review]. Det Nationale Forskningscenter for Velfærd, Copenhagen.

The Economic Council of Greenland. (2017). *The Greenlandic Economy 2017.*

Statistics Greenland. (2019). *Hovedbeskæftigelse blandt fastboende fordelt på tid, branche og uddannelse* [ARDBFB8]. http://bank.stat.gl/pxweb/da/Greenland/Greenland__AR__AR30/ARXBFB8.px/?rxid=a245454f-3113-4f62-a25f-2e0a5b2534d1

The Economic Council of Greenland. (2018). *The Greenlandic Economy 2018.*

The Economic Council of Greenland. (2019). *The Greenlandic Economy 2019.*

The National Bank of Denmark. (2019). *GRØNLANDSK ØKONOMI Stærk vækst og mangel på arbejdskraft* [GREENLANDIC ECONOMY Strong growth and shortage of labor]. Copenhagen.

8 Gender and the Greenlandic labour market

Helene Pristed Nielsen

Abstract

The Greenlandic labour market is highly gender segregated, and women and men tend to face different opportunities and have different experiences with labour market participation. In this chapter, we argue that we must understand some of the underlying mechanisms and reasons to approach an overall understanding of the Greenlandic labour market. Starting with Greenland's "female deficit" and the historic background explanations for this, the chapter presents existing knowledge on demographic developments in Greenland, their links to current gender roles, and the high levels of geographic mobility across the country. Proceeding to discuss gendered differences in educational attainment and segregation between the public versus private labour market, these overviews of current state of the art regarding gender and the Greenlandic labour market lead to a presentation of actual experiences with labour market participation among Greenlandic men and women. Based on ethnographic fieldwork stemming from two research projects on gender and labour market participation undertaken in South Greenland, Chapter 8 presents a handful of case stories that illuminate the trends discussed and provide grounds for further considerations of how best to tackle some of the gendered challenges facing the Greenlandic labour market.

8.1 Greenlandic gender relations

There are fewer women than men living in Greenland. Of the population, 53% is male and 47% is female (Høgedahl & Krogh, 2019, p. 11). This may seem like a simple fact, but the underlying reasons and potential future effects of this state of affairs are complex and multifaceted. It has to do with historic hardships of pure survival in a harsh climate, Danish interventions as colonising powers, current social policies, educational policies, global urbanisation trends, social expectations about men's and women's roles in society, and much more.

Historically, there has always been fluctuation in the ratio of women to men in Greenland, and as argued by Hamilton and Rasmussen (2010), these fluctuations have been closely linked to parameters such as climate, access to technology, and production conditions. Nevertheless, Danish interventions as colonial powers in Greenland had a significant and lasting impact on local gender relations, as well as the ratio of males to females in the country. Thus, the Danish so-called "modernisation policy" pursued in the 1950–1980s, according to several authors (Eistrup & Kahlig, 2005; Gaini, 2017; Poppel, 2005; Weyhe, 2011), partly laid the ground for current patterns of gender segregation in education and labour market participation. The Danish colonial rule aimed at creating a wage-earner mentality, which, according to Eistrup and Kahlig (2005), entailed two major changes to gender relations. Firstly, Greenlandic women were in paid employment in the fishing industry, while the men were fishing. Therefore, the female population was regarded as crucial to the modernisation policy. Secondly, the relationship between young female Greenlandic wage earners and young male Danish public servants and workers affected the role of native Greenlandic men (Eistrup & Kahlig, 2005, p. 204; see also Poppel, 2005). Weyhe (2011) also discusses how a large group among Greenlandic men were sidetracked during the modernisation policy, and he argues that this happened with reference to a perceived lack of discipline and qualifications. Furthermore, speaking the language of the colonial administration became a socially divisive factor. "Men who did not speak Danish found it difficult to find work and influence their own life situation, as well as the overall development in society" (Poppel, 2005, p. 136, own translation).[1] Adding insult to injury, Danish civil servants were endowed with particularly favourable pay and employment conditions in accordance with the birthplace criterion of 1964, which was not abolished until 1990 (Poppel, 2005, p. 133).

One source of the current female deficit in Greenland can be traced to the extensive immigration to Greenland from Denmark since the 1950s. This influx of workers from Denmark consisted mainly of men of working age who typically chose to settle in Greenland for a shorter period (Poppel, 2005). As argued by Gaini (2017):

> The foreign worker was assigned a symbolic role as nation builder in the colonial context. In the heyday of the modernization project, most foreign workers came to Greenland without their wives and children, thus often performing their masculinity in decadent style, something that would have been sanctioned as distasteful and improper behaviour in the workers' home countries.
>
> (pp. 57–58)

These Danish workers usually returned home, sometimes bringing with them their new Greenlandic wives (Hamilton & Rasmussen, 2010, p. 49). In this way, immigration from Denmark gave a boost to the gender imbalance, but also the emigration of women from Greenland (again, mainly to Denmark) has been significant. Actually, according to Hamilton and

Rasmussen (2010), most of the female deficit in Greenland can be explained by including the Greenlanders living in Denmark but born in Greenland in census counts.

Gaini (2017) further discusses how Greenlandic men were adversely affected by the Danish modernisation policy:

> Many workers [from Denmark] performed their masculinity as if there was no native masculinity of any sort in the colony. They did not recognize Inuit men as masculine men fitting their masculine modernization project in Greenland. In Greenland – most clearly from the 1960s and up to the 1980s – foreign (mostly Danish) workers held a powerful position that they as blue-collar workers could never have achieved at home. They shared a feeling of being totally free and invincible amongst Inuit. The demasculinization of Inuit men, together with the hypermasculinity of Danish men, displays a gendered expression of what could be called 'white supremacy'.
>
> (p. 58)

According to Gaini (2017), these historic events still have repercussive effects on current gender relation and masculinity ideals in contemporary Greenland. Poppel also argues that developments in the labour market have created major changes in gender relations, partly because educated and employed Greenlandic women are no longer as financially dependent on their husbands or partners as they have been in the past. Thus, "women are increasingly becoming the main providers" (Poppel, 2010, p. 55, own translation). Rasmussen (2009) further states that even successful fishermen and hunters in practice often rely on their spouses' incomes (see van Voorst, 2009). Hence, a growing number of households depend on incomes from the women, and Rasmussen (2009) reports that in more than half of the homes where hunting and fishing represent a significant part of the livelihood, it is women who earn most of the money. "Single men without these income sources, however, are confronted with severe economic problems" (Rasmussen, 2009, p. 526). An increasing number of these may often rely on living as "couch surfers" with family and friends.

8.2 Mobility in Greenland

In the context of labour market analyses, mobility may mean a number of different things (Høgedahl & Krogh, 2019, p. 50–51). Firstly, there is geographic mobility, which entails people moving between places. Being a vast country, with no roads between villages and settlements, geographic mobility is costly and difficult in Greenland. Nevertheless, geographic mobility is high, and the frequency with which people move and the distances across which they move are much higher than in the remaining Nordic countries (Nordregio, 2010). Secondly, labour market mobility may be understood as referring to a change in profession or an upskilling of workers. Finally, mobility may also refer to job mobility – the frequency of job turnovers.

With a very low population density, few jobs available in smaller settlements, and educational institutions dispersed across great distances, however, geographic mobility and changes in profession and/or upskilling tend to go hand in hand in Greenland. Furthermore, mobility patterns are gendered. Especially, the village population is geographically highly mobile; the young men from the villages often exhibit a mobility pattern that can be characterised as "bungy jumping" (Pristed Nielsen, 2016). This pattern entails that they, while closely linked to their communities and maintaining strong local anchorage, are simultaneously mobile, as they perform a series of consecutive temporary relocations. It is especially among men from low-income groups that it is possible to observe this pattern with many more or less short-term temporary employments. The young men are thus able to maintain connections and networks with their places of origin, while simultaneously exploiting income opportunities arising from fishing or extraction of raw materials, where there may be a possibility of either working on a fixed turn (working overtime away from home in periods combined with free time at home) or on more short-term project-related tasks (Nordregio, 2010).

Previously, more young men than women moved, but today the number is similar. However, there is a big difference *in the way they move*, and this difference results in a gradual decline in the number of women in smaller communities:

> As regards the women, there is a 'step-stone mobility pattern' in which they incrementally move from smaller to larger localities, and eventually choose to leave the country, while men on the other hand are more temporary movers, seeking jobs with shorter or longer duration in nearby towns and villages, but also more often return to where they came from.
>
> (Nordregio, 2010, p. 16, own translation)

Consequently, the female deficit varies from place to place, and "there is a clear correlation between gender distribution and economic activity, in the sense that more activities in the tertiary sector also mean a relatively higher proportion of women" (Nordregio, 2010, p. 62, own translation). A study from the late 1990s in Iceland identified a clear correlation between the population size in a municipality and the number of women relative to men: the smaller a community, the greater a loss of women (Hamilton & Otterstad, 1998, p. 15). However, the Icelandic study concluded that it is not the size of the community, but the nature of its labour market (i.e., the number of jobs in the tertiary sector and a diversified supply of jobs) that is critical to whether it is likely to encounter an imbalance in the number of male and female residents (Hamilton & Otterstad, 1998, p. 20). This finding seems directly transferrable to a Greenlandic context. Greenlanders with formal education move more frequently than those without, and those who move and acquire a higher education rarely move back to their original places of residence (Nordregio, 2010). Rafnsdóttir (2010), who writes more generally about the West Nordic region (Greenland, Iceland, and the Faroe Islands), explains that "especially young women from the outskirts who travel away experience it as difficult to move back home. The labour market is perceived as male, the educational opportunities as limited and women- and family-friendly welfare facilities as few" (p. 9, own translation).

This point may well be valid when considering social expectations in the smaller Greenlandic communities. Poppel (2010) emphasizes that it is important to keep in mind that the Greenlandic welfare model is decisively different from the Danish one, as a number of social services are means tested (p. 66), which in many cases makes women more dependent on their families than they are (or would be) in Denmark. Furthermore, a number of decisions concerning, for example, local housing policies and day care facilities have been devolved to district council decision-making. Such administrative patterns can play important roles in the decisions of young women to move elsewhere, because there generally are fewer women than men represented in political bodies in Greenland, and women are particularly underrepresented in village councils (Poppel & Kleist, 2009, p. 355).

Differences in the type and degree of mobility thus depend on place of origin and gender, as well as social expectations for gender roles. This is also underscored in the report from Nordregio (2010): "Greenlandic men, therefore, see their home base as a base they can leave for a shorter or longer period, while women regard it as an (individual and social) obligation to maintain this base function" (p. 17, own translation). Furthermore, van Voorst (2009) concludes based on a case study from South Greenland that moving from a small settlement to the local main town makes it easier for women to break with established gender roles inside their households.

Regarding labour market mobility (understood broadly as change in profession and/or job turnover) and flexibility (understood as the amount of hours worked and when), several authors point to a phenomenon that may be summarised as "romantic notions of freedom". For example, a recent ethnographic study in Narsaq, South Greenland, reports finding positively connoted ideas about freedom in interviews focusing on labour market experiences and refers to "the romantic liberty of the fishing and hunting community, where people sail or ride (dog sledge) to gather supplies when needed" (Pristed Nielsen et al., 2020, p. 77). This idea of freedom, as this study argues, "lingers throughout the interviews, and is understood as a positive thing in relation to deciding for oneself when to work" (Pristed Nielsen et al., 2020, pp. 77–78). This finding is supported by another study from Northwest Greenland, more precisely Upernavik. Here, Hansen and Tejsner (2016) undertook interviews with young male residents, inquiring into interests in potential jobs in the offshore oil industry in Baffin Bay. However, although development of a local oil industry could potentially create regular full-time employment, this is not necessarily seen as attractive to local young men:

> The young men expressed an interest in potentially supplementing their hunting activities with jobs in the oil industry. But it was clear that they first and foremost consider themselves hunters and only want to take other jobs if it is possible to use the opportunity to support the life they already live.
>
> (Hansen & Tejsner, 2016, p. 91)

Regular full-time work potentially leaves too little flexibility or freedom. In fact, according to Poppel (2005), "work" has historically been seen in Greenland as something for those unable to fend for themselves, such as women and *kifakker* (men unable to hunt and provide for themselves, who were thus dependent on the main hunter). By extension, Hansen and Tejsner (2016) point out that being "unemployed" in a Greenlandic context is not the same as not working: you may be hunting or fishing, even if not partaking in formal paid work. Work is not necessarily a matter of either/or, but perhaps of both/and. The same point may also be one of the underlying explanations for why Høgedahl and Krogh find a relatively large number of respondents who report themselves to be looking for jobs, but also report that they are unable to take up a new job within the coming two weeks (Høgedahl & Krogh, 2019, p. 23).

The role of hunting as social legitimation for being unable to take up a paid position, as well as the gendered connotations of this, is also evident from van Voorst's article (2009), which is tellingly entitled "'I work all the time-he just waits for the animals to come back': social impacts of climate changes: a Greenlandic case study". Although primarily concerned with how climate change impacts local gender relations, van Voorst also engages in a critical discussion of what "work" means in a gendered Greenlandic context. As she explains, hunting still carries a highly symbolic meaning, especially in the small communities, where sharing of prey also contributes to upholding social hierarchies. However, partly due to climate change and the resultant need for new and better equipment, "most hunter males have become financially dependent on their wife's income" (van Voorst, 2009, p. 243). Hence, "Greenlandic women in the settlement earned their own money, yet they were not able to control their incomes due to the high value attached to their husband's occupation" (van Voorst, 2009, p. 244).

8.3 Gender differences in education and labour market participation

Greenland is the Nordic country that spends the most public funding on education (Haagensen, 2014). Nevertheless, and despite prolonged efforts and national education plans, boys especially are "lagging behind" in educational attainment (Boolsen, 2013). Thus, "women in Greenland are generally better educated than men" (Faber et al., 2015, p. 72). This applies to secondary education, for example, 62.1% of the students who graduated between 1998/1999 and 2002/2003 were women (Eistrup & Kahlig, 2005, p. 210). When it comes to higher education, Poppel (2010) explains that women in particular have made use of new educational opportunities since the late 1980s (p. 49). Ilisimatusarfik/The University of Greenland was founded in 1987, and the proportion of women exceeds men in all degree programs offered at the university (Poppel, 2010, p. 53). According to Haagensen (2014), in 2013, women accounted for 68.3% of all graduates from higher educations in Greenland.

Furthermore, educational choices in Greenland are highly segregated, as there is a "general predominance of men in vocational training, and in turn, women dominate in both professional and higher educations" (Eistrup & Kahlig, 2005, p. 210, own translation). Boolsen (2010) further stresses that women in Greenland constitute "the educational elite". Adding to the significant correlation between gender and education, a correlation between being born in a town versus in a small village is evident when looking at education levels. Boolsen (2010, 2013) emphasizes how it is specifically youth from the towns who obtain an education, but village youths are lagging behind. Research by Rasmussen et al. (2011) confirms this pattern, and they stress that the educational system in the small villages makes it difficult for young people to succeed in school and especially proceed to higher education. Their results show that almost 50% more women than men complete a primary school exam in the small villages (Rasmussen et al., 2011, p. 94). Hayfield et al. (2016) report that, in 2013, almost twice as many women than men graduated from upper secondary level education (p. 50). Recent data confirm that young women are enrolled in education to a higher extent than young men are. However, at the same time, more young women younger than 30 are neither enrolled in education nor employed (Høgedahl & Krogh, 2019, p. 65). The latter, most probably, has to do with Greenlandic women tending to have their first child at a much younger age than women in the remaining Nordic countries do; hence, many of them are presumably on maternity leave.

In addition to the pronounced gendered difference in educational attainment in Greenland, there is a strongly gendered division between the public and the private sector, referred to in the literature as "horizontal gender segregation" (Bloksgaard, 2011). Hence, "public administration and service is by far the largest business sector in Greenland and accounts for almost 40% of all employment. ... [C]lose to two-thirds of all working women in Greenland and more than twice as many women than men work in this sector" (Hayfield et al., 2016, p. 46). It is true for all of the Nordic countries that women (Faber et al., 2015) dominate the public sector; however, the extent to which this is the case in Greenland is outstanding.

In contrast, the Greenlandic labour market is characterised by a *less* gendered division in part-time work than in other Nordic countries. Thus, in terms of part-time work, Greenland differs significantly from the remaining

Nordic region on several parameters. Firstly, very few people in Greenland are officially reported as working part time, and only 10% of respondents in the study by Høgedahl and Krogh (2019, p. 46) report themselves to be working part time. Secondly, "in Greenland, there are no gender differences in part-time work, yet a significant difference is found between people in rural and urban areas, as the former are more likely to work part time" (Hayfield et al., 2016, p. 9). In the remaining Nordic countries, gender is the strongest predictor for part-time work, but in Greenland, it is the urban–rural distinction that correlates most strongly with part-time work. Thus, "it is notable that women in urban areas are more likely than men in rural areas to hold full-time jobs" (Hayfield et al., 2016, p. 46). It would thus seem that the availability of full-time work in urban areas (and its relative unavailability in rural areas), coupled with the relative prominence of male inhabitants in rural areas, counters what is otherwise a frequently encountered pattern of gender segregation in part-time work.

When it comes to differences in wages, Greenlandic women are – despite their high labour market presence and education levels – underrepresented in managerial positions and have lower average earnings than Greenlandic men have. However, according to Poppel (2010), this pay gap has substantial local variations: the pay gap in the towns is far greater than in small villages (p. 63). Hayfield et al. (2016) find that

> Despite the small difference in part-time work between men and women, men earn significantly more on average per month than women. This is something of a paradox, since higher educational background correlates with higher wages and since women are better educated than men. It is evident, then, that other factors are also in play. The difference in earnings is likely to be more related to different employment types and typical 'female' jobs being lower paid.
>
> (p. 55)

Notwithstanding a shortage of labour within specific parts of the Greenlandic public sector, for example in teaching, it is difficult to detect any incentives for men to opt for "female" jobs. However, Weyhe (2011) concludes that this does not mean that attempts should not be made to focus on gender and diversity on Greenland's labour market. Hence, he argues that "challenging the gender stereotypes preventing individuals from moving freely across the gender segregated labour market remains essential" (Weyhe, 2011, p. 259, own translation).

8.4 Case stories about gender, place, and labour market participation

In this section, I present a handful of ethnographic case stories based on narratives collected among men and women who have participated in some of the research projects with which I was previously involved in South Greenland. The cases are provided as examples to illuminate some of the wider trends and themes discussed above and to provide real "flesh and blood" on some of the points raised about how and why a deeper understanding of the Greenlandic labour market necessitates insights into aspects such as culture, history, mobility, and gender roles.

I draw on data from the two projects "Sustainable business and demography: Exploring critical links between gender, youth and small-scale business development in fisheries and tourism in South Greenland" [SBD], led by my colleague Lill Rastad Bjørst (2017–2018), and "EQUIL – Equality in Isolated Labour Markets", led by myself (2018–2020). The SBD project comprises a data set consisting of 14 interviews with 22 people (some interviewed jointly) from South Greenland, collected on location in January 2018.[2] In addition, this data set includes impressions and information gathered during a workshop held in Qaqortoq in April 2018, where I participated personally, as well as information derived from my participation in various workshops in Denmark on business development and entrepreneurship in Greenland. Furthermore, I conducted five interviews with former residents of South Greenland, who at the time of interviewing lived either in Denmark or in Nuuk. The EQUIL project, on the other hand, was comparative between Denmark, the Faroe Islands, and Greenland. For the purposes of the present chapter, I draw solely on the Greenlandic data set, consisting of 10 interviews with females living in and around Narsaq, collected during April 2019.[3] Additionally, I draw on impressions and recorded discussions from a workshop held in Tórshavn on the Faroe Islands in May 2019, bringing together informants from all three of our data collection sites to discuss business development, gender relations, and prospects for boosting settlement in their respective locations, which were all characterised by dwindling population figures and a notable female deficit.

Below, I present four case stories, retold by me, but based on working life experiences as recounted during interviews, workshops, and debates among informants in or from South Greenland. I have renamed all participants, and unless marked in quotation marks, the stories represent my summary or paraphrase of what they said. The stories serve to make the more general observations more concrete and give further insight into the lived consequences of contemporary opportunities and challenges faced by women and men in accessing the Greenlandic labour market.

Ivalu was a young woman in her 20s living on a sheep rearing station with her husband and two small children. She originally wanted to train as a police officer in Nuuk, but then her father-in-law died, and her husband inherited the farm. Ivalu was only 22 years old when this happened. Consequently, they abandoned their Nuuk life and her city aspirations: "It's a tradition that the oldest son takes over the father's farm [...] this is something which you feel in your heart". She now lives as a farmer's wife, participating in the hard work on the farm. She spoke about a challenging life and needing to be able to fend for oneself, for example when the water pump froze to ice, the tractor broke down, or the sheep were ill. As she said, "You need to be able to do almost anything. You can't just call the mechanic". Besides her work on the farm, she earned money on the side by catering to tourists during the summer, for example offering bed and breakfast and horseback rides. Furthermore, she worked in a nearby village, mainly during winter. The latter job caused her challenges, however, because ice conditions were deteriorating. Thus, she had previously been able to rely on making the trip to work via a 30-minute snowmobile ride across the ice on the fjord. However, as of late, she had to take the much more tedious and difficult 2.5-hour route via the dirt track road on her ATV.

Ivalu was a tough young woman. She remarked, "You can't just sit down and cry even if you feel like it [...] well, maybe you do anyway, but once you are done crying, you think 'well, nobody will come and help you, so you had better ...' You learn this the hard way". In addition, she was also a modern woman. She advertised her bed and breakfast offers on the internet and maintained ties with former guests through Facebook. Some guests, however, seemed too modern for her taste. She explained how many American tourists would require vegetarian diets – if they did so based on a concern for climate impact, they had much rather eat her homemade lamb or fish dishes than request vegetables sailed or even flown in from abroad. She was content with her life and expressed that she felt enriched simply be being alive and present in her place. "I do not have to look at a watch, I do not need to listen to people, or think about what they have said. I'll just be here and what you see is just really beautiful and it's pure nature".

Only one among several women with whom we talked, Ivalu managed to arrange her private and her working life to her satisfaction. Another woman, old enough to be Ivalu's mother, spoke about a life of multiple job changes, which had increasingly brought her closer to her dream of living an independent life in harmony with her natural surroundings. This woman, Nivi, stood out from the general mobility pattern by living in the exact same house in which she was born and raised. In this sense, she had exhibited very little mobility during her life. Her work–life mobility, however, was exceedingly high. Just within the last decade or so, she had undertaken various income-generating activities, including working as a hairdresser, internet café owner, coffee shop vendor, taxi driver (until someone ran into her car and she couldn't find a mechanic to repair it), and entrepreneur. Her entrepreneurial efforts consisted in setting up production of organic herbs, teas, and skin products. In fact, she had been sufficiently successful as to allow herself to concentrate 100% on this business, even being able to hire staff for production and packaging her goods, which were approved for export to a number of countries.

A somewhat reverse story could be told of Hans. Maybe a little older than Nivi, he had always been able to rely on an income as fisherman through selling his catch to the local fish factory. However, the factory had closed a few years ago, and any fish he caught, he either had to rely on selling directly to private consumers, or eat himself – in this sense relying on subsistence economy. "But I have six full sized freezers and no more room for meat or fish!" he exclaimed. With food on the table (including for family and friends) and a house and boat already paid for, he had of late been prone to simply stacking his fish crates on the harbour front and asking the local boys hanging around at the harbour to spread the rumour that people could come and pick up what they needed. A new somewhat unanticipated source of income had turned up, however. This consisted in Hans taking tourists on "adventure" during the summer – this entailed simply taking them on a short boat trip around the fjord, fishing trips, or even whale watching. However, many restrictions surround these activities, for example related to safety regulations, boat licenses, and (and most troubling) the fact that if he earned more from tourism than fishing, he would lose his fishing licence. If it was not for his pride in his profession and the local scenery, Hans was not quite sure the income he earned in this way was worth the effort.

Minik, on the other hand, was a young man taking immense pride in his professional interaction with tourists. Being educated as professional tourist guide, he spoke about receiving international tourists from Germany, Spain, and the United States. While most of these were simply passing through, they also left lasting impressions locally, for example on Minik, who felt his life was being enriched by working with foreign travel agents and meeting people from all corners of the world. He also spoke about how locals reacted to him leading tourists through town on his guided tours. "I meet many different inhabitants here in town who support me and greet me and say a word, [...] a slang word they say to me in Greenlandic, which means 'how good you are' or, like, recognising me for speaking English and explaining things to tourists". Being recognised for his formal education and abilities, as well as his embodied knowledge of the local nature and climate, Minik openly acknowledged the value of further education, although prizing his current level of freedom.

> I live in the present. In my generation, especially me, on my family's side, I decide for myself when to go to get education or work and how to live. My mom supports me no matter what I decide. She just wants to make sure I have an education, because then I secure the future. You can say that if you work with basic education and for unskilled salary, it is not as much as a skilled worker. It is completely different. That is why I have to have an education. I can't imagine what education I want to take. It is a little complicated for me to think about the future.

Having not yet made up his mind exactly about which further education to pursue, Minik was certain, however, that he would get some sort of education, if nothing else, to gain access to higher income levels. The choice of education was inextricably linked with a decision about where to move to achieve this education. Minik felt this decision was slightly challenging.

Where would he live, and would he be ok living far away from family and friends?

8.5 Discussion

None of the people described above held regular full-time wage employment based on the Greenlandic standard of a 40-hour workweek. This does not mean that such employment does not exist; obviously, it does (Høgedahl & Krogh, 2019). It is partially a reflection of the fact that both of the ethnographic field studies, from which the stories come, focus on entrepreneurship, small-scale business development, and gendered labour market participation. Nevertheless, I argue that it is also a reflection of some wider lessons to learn about gender and the Greenlandic labour market.

Firstly, I argue that it is significant that the working lives described above are characterised by a pattern of pooling income from a variety of sources and somewhat opportunistically changing direction or making most of the available resources whenever possible. This point is in line with arguments by Dahlström et al. (2006) about "how to make a living in insular labour markets". They address the research question "How do people generate income in insular areas, where a daily commute to a neighbouring labour market is unrealistic?" (Dahlström et al., 2006, p. 9), focusing on six different insular areas within the Nordic region. Although not including Greenland among their case studies, many of their arguments seem transferable to this context. As Dahlström et al. (2006) point out, it is necessary to move beyond the framework of a traditional employee/employer understanding and the idea of lifelong wage work, in order to understand the dynamics of insular labour markets. Thus, they argue, "In insular labor markets, multiple job holding is particularly common, and includes transitions both within the labor market and between the labor market and other support systems such as unemployment benefit" (Dahlström et al., 2006, p. 13).

Thus, the authors argue for an expanded notion of "income", including welfare benefits and the "black" or "informal" economy – that is, the part of the income not disclosed to the tax authorities, which they argue plays a potentially important role in insular labour markets (Dahlström et al., 2006, p. 9). Although none of our interviewees and workshop participants spoke directly about "black economy", there were plenty of examples about sharing available resources and/or swapping favours with friends and neighbours. The most obvious example was Hans sharing his catch, but at times Nivi had sold cakes baked by friends in her former coffee shop and sheep farmers swapped meat for other goods, such as honey.

Meert (2000) also addresses "alternative" sources of household income in rural areas, albeit in a Belgian context. He includes "redistribution" (for example in the form of charity and welfare) and "reciprocity" (in the form of mutual favours and sharing) together with "remunerated activities" in understanding economic integration of households in rural communities

(Meert, 2000, p. 320). Although developed in a different geographic and demographic context, Meert's (2000) points are relevant in recognising the need for an expanded notion of income in a Greenlandic context. Based on a fieldtrip to northern settlements and interviews with fishers and hunters, Hansen (2018) observes, "in the northernmost settlements, hunting is dominant and most live on catch alone. A large proportion of the catch contributes to local households and is part of a subsistence economy" (p. 4, own translation). Furthermore, van Voorst (2009) points out how local villagers participate in a "shared adaptive strategy of food sharing" (p. 241), a practice which her informants tell her has been on the rise in recent years, due to difficulties partly related to climate change and, therefore, changing conditions for hunting and fishing. Hence, standard models of measuring household income levels and employment levels are not necessarily easily applicable in a Greenlandic context, certainly not in the smaller villages, where much undocumented work directly and indirectly contributes to local household economy and a culture of sharing prevails (Hansen, 2018; van Voorst, 2009). Hence, Hansen (2018) also describes how "fishing and hunting are not perceived as a profession, but rather as a foundation for, and an integral part of, people's everyday lives" (p. 5).

Secondly, I argue that, short of gender determinism, some of the differences between the experiences of the men and the women in my case stories are symptomatic of wider trends in Greenlandic society. Above, I reference a point from the Nordregio report (2010) about how women frequently see it as their responsibility to maintain base or nest functions. In different ways, this seems to be the case for Ivalu (maintaining her husband's original base, now the base of her own children) and Nivi (contributing to the town of her upbringing and fighting to make other avenues of income available than the extraction of minerals). Minik, however, also exhibits close relations to his "base", exemplified in how considerations about distance are a factor in his thoughts about choosing a line of education.

To some extent, the experiences of Hans over his lifetime also echo points raised by Gaini (2017) and Weyhe (2011) about how traditional fisher and hunter masculinities are under pressure because of changes in the labour market structure. Hans, of course, makes his own adaptations, and the ways in which he and Minik put their knowledge about local nature and wildlife to use in the budding tourism industry bear witness to how traditionally "male knowledge" may find new outlets on the Greenlandic labour market. Nevertheless, other authors also trace evidence of changes in gendered patterns on the Greenlandic labour market, for example Hayfield et al. (2016) remark,

> The fact that women receive almost the entire parental leave payment strongly indicates that they take on more of the total care responsibility in households with children. However, research on male and female perceptions of their contributions to the family suggests that gender roles

in Greenland are changing and that more and more females now have the main income in Greenlandic households.

(p. 55)

Furthermore, research evidence suggests that gendered patterns of job segregation differ between villages and towns: "In contrast to men living in small villages, far more men living in towns show an interest in jobs in health, social institutions and teaching. This difference both reflects a difference in attitudes and a difference in which types of jobs are *de facto* available in these different areas" (Nordregio, 2010, p. 19, own translation).

Third, I argue that the income-generating activities and the occurrence of new opportunities (as well as the closing of opportunities) are closely related to the characteristics of South Greenland as a place. For researchers occupied with rural development, it has long been apparent that place matters, also in understanding predicaments for income generation (see, e.g., Lindsay et al., 2005; Meert, 2000; Robertson et al., 2008). Farole (2013) observes how "economic activity is unevenly distributed across places" (p. 15), with a general tendency to a widening gap between rural and urban places. The DORA model (which is an acronym for "Dynamics of Rural Areas"; Bryden & Hart, 2004) attempts to account for uneven developments in various rural areas across Europe, recognising the interdependencies of economic factors and questions of social equity (Bryden, 2011, p. 7). The main point of this model is that several tangible and intangible factors play a role in local labour markets and that the intangible factors often interact with the tangible ones. Thus, the DORA project "identified the important local influences of institutional performance, culture and community, and quality of life on more tangible factors such as natural resources, human resources, infrastructure, investment, and economic structures" (Bryden, 2011, p. 10).

Therefore, opportunities for income generation are place specific, and contextual analyses, including sociological perspectives, are necessary in understanding local developments. Concretely, in South Greenland, place-specific opportunities manifest themselves, for example in farming; South Greenland is the only municipality in Greenland where the climate is mild enough to support such types of income-generating activity. This impacts Ivalu's working life and choice of where to settle, but also provides Nivi with arguments for expanding local reliance on natural botanic resources rather than inviting in mining companies. Farming is not only a place-specific opportunity in South Greenland but also, within certain segments of the population, a socially respected mode of providing for oneself, associated with quality of life and respect for nature and tradition by several of our informants.

Bryden and Hart (2004) do not explicitly include aspects such as gender roles or colonial history among their intangible factors, but based on the arguments presented above, understanding the significance of both these factors seem highly conducive to developing a deeper understanding of the Greenlandic labour market. For example, although Minik relies on a new source of income, namely tourism, the recognition he receives for his skills and the way in which he activates traditional "male" knowledge about nature and how to navigate it may be seen as linked to traditional gender roles and modes of recognition for contributing to the community. Furthermore, he specifically refers to the praise attached to his English skills – in this case, more relevant than Danish skills, which is otherwise the most important second language in Greenland (Langgård, 2005). It is notable that, in the interviews, Minik and Ivalu emphasise their communication with American tourists. According to Langgård, the current existence of language dilemmas in the Greenlandic labour market may act as a barrier for further development of the knowledge economy and labour market. As Langgård points out, for certain Greenlandic speaking segments of the population, it may be difficult to "distinguish the postcolonial position of Danish, and then Danish (together with English) as a means to procure information. Emotionally, the situation becomes more complicated by the need to also accommodate Danish-only speaking Greenlanders" (Langgård, 2005, p. 154).

In terms of how best to tackle challenges facing the Greenlandic labour market, any development of future policies would most likely benefit from taking gender, post-colonial, and language aspects into account. Together with other factors identified by Bryden and Hart (2004) as more "tangible" – including issues related to infrastructure, natural, and human resources – the "intangible" factors add depth to our understandings of why decades of heavy investment in infrastructure and education seem insufficient in breaking existing barriers within the Greenlandic labour market. Including "intangible" factors in policy development might even lead us to ask other questions about the purpose of such policy development: for example, whose interests are being furthered if working equals earning wages or partaking in a 40-hour working week standard?

8.6 Conclusion

Pulling the threads together on the above discussions and the presentation of current state of the art regarding gender and the Greenlandic labour market, two overarching points stand out. Firstly, it seems difficult to speak of one Greenlandic labour market. Everything from income-earning opportunities (whether pecuniary or subsistence based) to social expectations and gender roles seem to differ between towns and smaller settlements. Furthermore, as commuting is an impossibility in Greenland, changing jobs often involves geographic mobility, maybe for the entire family. Therefore, even if attractive jobs are available elsewhere (typically in a larger location than from where one comes), making the transition to town life may change family life on several parameters and, therefore, be a difficult transition to make. Secondly, although the heavy horizontal gender segregation (between public and private sectors) of the Greenlandic labour market is certainly not unique when compared to its Nordic neighbours, the vertical gender segregation (between high- and low-paid jobs) is not only gendered but also partly ethnically based, with more ethnic Danes at the top rung of public and private management. Given historical experiences of labour market developments in Greenland, some may feel this is the bigger fish to fry before moving on to the question of gender equality.

Notes

1. This point may be equally valid for women, but Poppel's focus in this publication is specifically on "Barriers for Greenlandic men in the labour market".
2. Thanks are due to Rikke Becker Jakobsen, Aalborg University, who collected the interviews during a two-week field trip to Qaqortoq and neighbouring towns and settlements during January 2018. Furthermore, thanks to research assistants Anna Stegger Gemzøe and Stine Als Brix for transcribing the data set, as well as conducting interviews in Denmark.
3. Thanks are due to Steven Arnfjord and Karen-Marie Ravn Poulsen, Ilisimatusarfik/University of Greenland, who collected the interviews.

References

Bloksgaard, L. (2011). Masculinities, femininities and work—the horizontal gender segregation in the Danish labour market. *Nordic Journal of Working Life Studies, 1*(2), 5–21.

Boolsen, M. W. (2010). *Uddannelsesplanen: Rapport 3.* Nuuk: Det Grønlandske Uddannelsesdepartementet – Det Grønlandske Selvstyre.

Boolsen, M. W. (2013). Evaluating education in Greenland. How is power exercised through evaluation models? *Scandinavian Journal of Public Administration, 16*(3), 65–82.

Bryden, J., & Hart, K. (2004). *A new approach to rural development in Europe: Germany, Greece, Scotland and Sweden.* Lewiston, NY: E. Mellen Press.

Bryden, J. (2011). *Rural development indicators and diversity in the European Union.* Research Gate. https://www.researchgate.net/profile/John_Bryden3/publication/228865950_Rural_Development_Indicators_and_Diversity_in_the_European_Union/links/0046352f9d1cef2cb8000000/Rural-Development-Indicators-and-Diversity-in-the-European-Union.pdf

Dahlström, M., Aldea-Partanen, A., Fellman, K., Hedin, S., Javakhishvili-Larsen, N., Jóhannesson, H., Manniche, J., Olsen, G. M., & Petersen, T. (2006). *How to make a living in insular areas – six Nordic cases.* Stockholm: Nordregio.

Eistrup, J., & Kahlig, W. (2005). Historiske forandringer på ligestillingsområdet i Grønland som faktor for velfærdsudvikling' [Historical changes in the field of gender equality in Greenland as a factor for welfare development]. In A. K. Berglund, S. Johansson, & I. Molina (Eds.), *Med periferien i sentrum: – en studie av lokal velferd, arbeidsmarked og kjønnsrelasjoner i den nordiske periferien* (pp. 201–214). Alta: Norut NIBR Finnmark.

Faber, S. T., Pristed Nielsen, H., & Bennike, K. B. (2015). *Place, (in)equality and gender: A mapping of challenges and best practices in relation to gender, education and population flows in Nordic peripheral areas.* Copenhagen: Nordic Council of Ministers, TemaNord 2015: 558.

Farole, T. (2013). *The internal geography of trade. lagging regions and global markets.* Washington: The World Bank.

Gaini, F. (2017). Crack in the ice: Marginalization of young men in contemporary urban Greenland. In C. Haywood & T. Johansson (Eds.), *Marginalized masculinities: Contexts, continuities and change,* New York: Routledge.

Haagensen, K. M. (2014). *Nordic Statistical Yearbook 2014.* Copenhagen: Statistics Denmark – Nordic Council of Ministers.

Hamilton, L. C., & Otterstad, O. (1998). Sex ratio and community size: Notes from the Northern Atlantic. *Population and Environment, 20*(1), 11–22.

Hamilton, L. C., & Rasmussen, R. O. (2010). Population, sex ratios and development in Greenland. *Arctic, 63*(1), 43–52.

Hansen, A. M. (2018). *Lokale holdninger til fiskeri, bæredygtighed og fremtid i Nordvest Grønland* [Local attitudes to fisheries, sustainability and the future of Northwest Greenland]. Aalborg: Aalborg Universitet.

Hansen, A. M., & Tejsner, P. (2016). Challenges and opportunities for residents in the Upernavik District while oil companies are making a first entrance in Baffin Bay. *Arctic Anthropology, 53*(1), 84–94.

Hayfield, E. A., Olavson, R., & Patursson, L. (2016). *Part-time work in the Nordic region III: An introductory study of the Faroe Islands, Greenland and Åland Islands.* Copenhagen: Nordic Council of Ministers. TemaNord 2016: 518.

Høgedahl, L., & Krogh, C. (2019). Den grønlandske arbejdskraftsundersøgelse (G-AKU) [The Greenlandic Labour Force Survey (G-AKU)]. Aalborg University, Center for Arbejdsmarkedsforskning (CARMA).

Langgård, K. (2005). Sproglige dilemmaer på det grønlandske arbejdsmarked [Language dilemmas in the Greenlandic labor market]. In A. V. Carlsen (Ed.), *Arbejdsmarkedet i Grønland – fortid, nutid og fremtid* (pp. 141–159). Nuuk: Ilisimatusarfik.

Lindsay, C., Greig, M., & McQuaid, R. W. (2005). Alternative job search strategies in remote rural and peri-urban labour markets: The role of social networks. *Sociologia Ruralis, 45*(1–2), 53–70.

Meert, H. (2000). Rural community life and the importance of reciprocal survival strategies. *Sociologia Ruralis, 40*(3), 319–338.

Nordregio (2010). *Mobilitet i Grønland: Sammenfattende analyse* [Mobility in Greenland: Summary analysis]. Stockholm: Nordic Council of Ministers Research Programme.

Poppel, M. (2005). Barrierer for grønlandske mænd på arbejdsmarkedet [Barriers for Greenlandic men in the labor market]. In A. V. Carlsen (Ed.), *Arbejdsmarkedet i Grønland – fortid, nutid og fremtid* (pp. 125–140). Nuuk: Ilisimatusarfik.

Poppel, M. (2010). Kvinder og Velfærd i Grønland [Women and Welfare in Greenland]. In G. L. Rafnsdóttir (Ed.), *Kvinder og Velfærd i Vestnorden* (pp. 38–68). København: Nordisk Ministerråd. TemaNord 2021: 578.

Poppel, M., & Kleist, J. C. (2009). Køn og magt i politik og erhvervsliv i Grønland [Gender and power in politics and business in Greenland] In K. Niskanen, & A. Nyberg (Eds.), *Kön och makt i Norden. Del 1: Landsrapporter* (pp. 341–358). København: Nordisk Ministerråd. TemaNord 2009: 569.

Pristed Nielsen, H. (2016). Offshore but on track? Hypermobile and hyperflexible working lives. *Community, Work & Family, 19*(5), 538–553.

Pristed Nielsen, H., Hayfield, E. A., & Arnfjord, S. (2020). *Equality in isolated labour markets. Equal opportunities for men and women in geographically isolated labour markets in Læsø (DK), Suðuroy (FO), and Narsaq (GL).* Copenhagen: Nordic Council of Ministers. TemaNord 2020: 522.

Rafnsdóttir, G. L. (Ed.). (2010). *Kvinder og Velfærd i Vestnorden* (578). København: Nordisk Ministerråd. TemaNord 2010: 578.

Rasmussen, R. O. (2009). Gender and generation: Perspectives on ongoing social and environmental changes in the Arctic. *Signs, 34*(3), 524–532.

Rasmussen, R. O., Roto, J., Olsen, L. S., & Harbo, L. G. (2011). *Status for bosteder i Grønland med særlig fokus på bygderne [Status of housing in Greenland with special focus on the settlements].* Stockholm: Nordregio.

Robertson, N., Perkins, H. C., & Taylor, N. (2008). Multiple job holding: Interpreting economic, labour market and social change in rural communities. *Sociologia Ruralis, 48*(4), 331–350.

Van Voorst, R. (2009). "I work all the time - he just waits for the animals to come back": Social impacts of climate changes: A Greenlandic case study. *Jàmbá: Journal of Disaster Risk Studies, 2*(3), 235–252.

Weyhe, T. (2011). Finanskrisen og arbejdsmarkedet i Grønland [The financial crisis and the labor market in Greenland]. In S. B. Nielsen (Ed.), *Nordiske mænd til omsorgsarbejde!: – en forskningsbaseret erfaringsopsamling på initiativer til at rekruttere, uddanne og fastholde mænd efter finanskrisen* (pp. 248–259). Roskilde: VELPRO – Center for Velfærd, Profession og Hverdagsliv.

9 NEETs and disadvantaged groups not in employment in Greenland

A national and international perspective

Rasmus Lind Ravn

Abstract

Youth unemployment is a contentious subject in Greenland, and the Greenlandic government is particularly concerned with reducing the number of young Greenlanders who are not in employment or education. Using survey data from the Greenlandic Labour Force Survey (LFS), we turn our attention to NEETs (Not in Education, Employment, or Training) in Greenland and focus on young people aged 17–29. First, we take an international outlook to explore whether the NEET challenge is greater in Greenland than in other selected countries. Then, we explore how NEETs differ from their young counterparts in employment or in education. Furthermore, we compare NEETS to the older group who are also not in employment to explore potential differences between groups. We discuss educational opportunities in Greenland, geographic mobility, and poor health as significant barriers for labour market and educational participation. The chapter ends by outlining and discussing Greenlandic policy initiatives to increase employment and participation in education.

9.1 Who are the NEETs and what are the implications of not being in education or employment?

Youth unemployment is high on the Greenlandic political agenda and is a contentious subject. One of the main priorities in the official Greenlandic employment strategy is a reduction of youth unemployment (Government of Greenland, 2014; Government of Greenland, 2015; Greenlandic Ministry of Finance, 2019) and increasing the educational level of the population is pivotal for the Greenlandic economy (Arnfjord, 2016; The Economic Council of Greenland, 2018, 2019). We outline and discuss the policies in place to address unemployment and inactivity later in the chapter. The present chapter focuses on Greenlanders who are not in employment, education, or training (i.e., NEETs). Exploring NEETs in Greenland is particularly relevant because the Greenlandic NEET rate is considerably higher than the rate in the remaining Nordic countries, as is evident from the chapter.

Originally coined in the United Kingdom in the 1980s, NEET refers to "Not in Education, Employment, or Training" (Yates et al., 2011, p. 514). Since then, the term has gained prominence and usage internationally to describe young people who are not in employment or participating in education. Whom to include in the NEET category is a matter of discussion. Originally, it only encompassed young people below the age of 18 (Maguire, 2015) but has since broadened to include 15- to 24-year-olds (e.g., Eurofound, 2017), 15- to 29-year olds (OECD, 2016), or even 15- to 34-year-olds, as is occasionally the case with the Eurostat's use of the term. Therefore, which age groups to include in the NEET category is not a clear-cut case but generally encompasses rather young people and not the entire working age population.

In general, NEETs have difficulties transitioning from school into further education or into employment (Henderson et al., 2016, p. 138; Maguire & Thompson, 2007, p. 1; VIVE, 2019a). Successful transitions from one of these phases into another is extremely important because periods of inactivity and unemployment increases the risk of isolation from society, poverty, crime, and poor health (Carcillo et al., 2015). Furthermore, periods of inactivity have long-term negative effects on future employment prospects and earnings, so-called scar-effects (OECD, 2015, 2016). From a governmental point of view, a large share of unemployed in a country is a costly affair, leading to high public expenditures on income-transferring benefits.

Reducing the number of NEETs is not an easy task, and there is no evident policy solution. Existing research shows that NEETs are disadvantaged in a number of ways and have poor preconditions for obtaining employment or completing an education. They often have poorer health than their young counterparts do and more often have unemployed parents or parents with low educational attainment (Carcillo et al., 2015). NEETs also often have low educational attainment. In the OECD countries, 85% of NEETs have only completed upper-secondary school as their highest educational attainment (Carcillo et al., 2015, p. 7). Furthermore, NEETs more often suffer from mental health problems (Carcillo et al., 2015). It is, however, difficult to distinguish between causes and effects. For instance, do mental health issues arise because of unemployment or inactivity, or do mental health issues prevent employment and participation in education? Evidence suggests both are the case, and they may be mutually reinforcing (Henderson et al., 2016).

As shown above, a great deal of research has studied what characterizes NEETs internationally, and comparative statistics are widely available. This is, however, not the case for Greenland. What is the typical educational attainment among Greenlandic NEETs, and which barriers do Greenlandic NEETs face? These are the types of question I seek to answer in this chapter.

Therefore, the aim and scope of the chapter is to provide some descriptive insights into what characterizes NEETs aged 17–29 years in Greenland. Do they differ from their young counterparts who are participating in education or are in employment?

Furthermore, we compare NEETs (17–29 years) to an older group (30–64 years) who are also not in employment or participating in education. The comparison between the young and the older group is relevant for a number of reasons. First, we know that unemployment and inactivity for all age groups can have detrimental consequences for the individual in the form of poverty, stigma, social isolation, and mental health problems (Gallie et al., 2003; Paul & Moser, 2009; Staiger et al., 2018). Second, income-transferring benefits put a strain on public expenditure regardless of the age of the recipient. Third, if young Greenlanders have lower employment rates than their older counterparts do, the scar effects of unemployment and inactivity for the young Greenlanders can have severe negative consequences for the Greenlandic economy in the long term.

As in Chapters 5 and 7, the analysis in this chapter is based on data from the Greenlandic Labour Force Survey (LFS) of 2019, the first of its kind conducted in Greenland (see Chapter 5 for an exhaustive description of the Greenlandic LFS). In addition to the LFS data, we utilize data from Eurostat to make cross-country comparisons. We use data from Statistics Greenland to supplement the LFS data and to compare the survey results to administrative data on employment and educational outcomes in Greenland.

The chapter is structured as follows. The first part of the analysis takes an international perspective to investigate whether the Greenlandic NEET challenge is smaller or greater than in other countries and compares rates of participation in education and educational attainment between Greenland and selected countries.

The second part of the chapter uses a series of descriptive statistics from the Greenlandic LFS to show what characterises Greenlandic NEETs and the older group not in employment or education. Furthermore, we explore self-reported explanations for not looking for work, and we report how NEETs characterize themselves. Then, we discuss barriers for employment and participation in education by drawing upon existing research. We outline Greenlandic policy initiatives to increase employment and to reduce the number of NEETs and relate these to policy initiatives in Denmark. We end the chapter by concluding and discussing the findings.

9.2 Greenlandic youth outcomes in an international perspective

In this section, we start out by taking an international perspective to investigate if Greenland faces a larger challenge in relation to NEETs and youth employment compared to selected European countries, especially the Nordic countries. Then, we compare the rates of participation in education and educational attainment between countries. We particularly focus on these challenges in relation to other Nordic countries and, in particular, Denmark.

We chose the comparison with the Nordic countries for the following reasons. The Nordic countries are internationally renowned for their macroeconomic performance and high employment rates among females and males. However, Nordic labour market research has somewhat neglected Greenland. Furthermore, a comparison with Denmark is crucial due to the colonial history, with Denmark as a colonial power in Greenland (Dahl, 2010; Rud, 2019; Thisted, 2016). The colonial history of Greenland and Denmark is relevant in this regard because existing research finds that colonization has a negative impact on economic development (Heldring & Robinson, 2012). In addition, Iceland, an additional Nordic country, is a former colony of Denmark located in the Atlantic Ocean, similar to Greenland, which makes comparison between Greenland and Iceland especially interesting. Iceland is also an island heavily dependent on fishing and its downstream industries.

In the comparisons, we furthermore compare the Greenlandic outcomes to those of three Southern European countries: Italy, Spain, and Greece. We included these countries because they are the most similar to Greenland in terms of the NEET rate.

9.2.1 NEETs and youth employment – the paradox of many NEETs but high youth employment

In the following, we explore the NEET rate and the youth employment rate to assess the extent of the challenges Greenland faces in relation to these two outcomes. First, we compare the NEET rate in selected European countries in Figure 9.1.

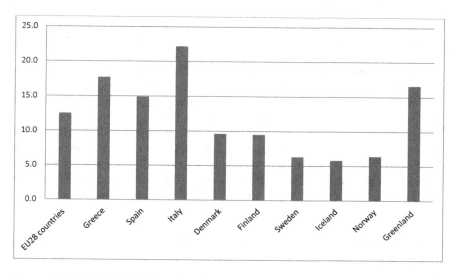

Figure 9.1 The LFS NEET rate in selected countries 2019 (ages 15–29 years).

Source: Eurostat [edat_lfse_21] and The Greenlandic Labour Force Survey.

Note: The Greenlandic NEET rate is based on a single survey whereas the rate for the other countries is based on the annual average.

We should interpret Figure 9.1 with some caution. First, the group categorized as NEETs in the Greenlandic LFS is quite small (47 respondents), which entails some statistical uncertainty. Second, the Greenlandic LFS did not ask the relevant questions concerning participation in education during the last four weeks, which reduces comparability across countries to some extent. Nevertheless, Figure 9.1 provides an indication of the challenges Greenland faces in relation to NEETs compared to other countries.

Figure 9.1 shows that the NEET rate in Greenland amounts to 16.6% at the time of the survey. This is well above the EU28 average of 12.5%. In this regard, Greenland seems most comparable to the southern European countries (Italy, Spain, and Greece) severely affected by the financial crisis of 2007–2008 that experienced a recession from which they have not yet fully recovered (OECD, 2016, p. 16). As such, it would seem that Greenland's challenges in this respect are quite large. This is further supported when we inspect the NEET rates of the other Nordic countries. With NEET rates of 9.6% and 9.5%, Denmark and Finland, respectively, have the highest NEET rates in the Nordic countries after Greenland. A difference of seven percentage points between Denmark and Greenland is a vast difference, especially when considering that Greenland is an autonomous country within the Kingdom of Denmark. The Greenlandic NEET rate is also more than 10 percentage points higher than that of Iceland, despite both countries being islands with small populations located in the North Atlantic Ocean and former colonies of Denmark.

However, looking at the youth employment rates alleviates this rather negative characterisation. Greenland stands out internationally with a very high youth employment rate, as depicted in Figure 9.2.

Figure 9.2 shows that Greenland has an exceptionally high youth employment rate of 69.9%, only surpassed by Iceland with a youth employment rate of 77.1%. In terms of the youth employment rate, Greenland is not at all like the Southern European countries. The LFS youth employment rate is roughly twice as high as that of the Southern European countries. Denmark and Norway have the third and fourth highest youth employment rates with 61.2% and 61.1%, respectively, which is well below the Greenlandic level. In addition to the timing of the data collection (the Greenlandic employment rate peaks in the spring and summer), the low share of Greenlandic youth participating in education can possibly explain the high employment rate, as explored in Figure 9.3 in the next section.

An apparent paradox thus exists because the Greenlandic NEET rate is high in comparison to other countries, and the youth employment rate is exceptionally high. The explanation for this lies with the fact that many young Greenlanders are not participating in education (see Figure 9.3) and are not actively looking for a job, which means that they are not counted as unemployed in the LFS unemployment definition.

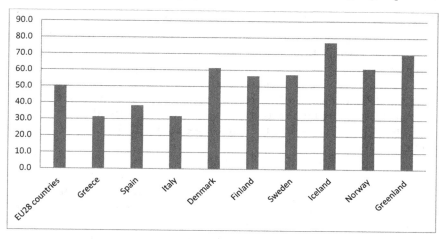

Figure 9.2 LFS Youth employment rate 2019 (ages 15–29).

Source: Eurostat [yth_empl_020] and The Greenlandic Labour Force Survey.

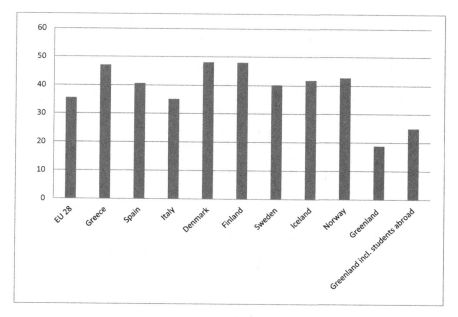

Figure 9.3 Share of 17- to 29-year-olds participating in education (upper secondary and above) (2018/2019).

Source: Own calculations based on Eurostat [educ_uoe_enra02 & demo_pjan] and Statistics Greenland [UDDISC11B & BEDCALC].

9.2.2 Low participation in education and many unskilled adults

As hinted above, only a small share of Greenlandic youth participate in the educational system. Educational opportunities are limited in Greenland, and many young Greenlanders therefore have to go to Denmark to study if they want to complete further education. In addition, young Greenlanders from small settlements (often rather remote and small villages) have to move away from home as rather young teenagers to attend upper secondary school in either Nuuk (the capitol), Qaqortoq, Aasiaat, or Sisimiut (larger towns).

Greenland is furthermore characterized by extremely high mobility – much higher than other Arctic countries and the Nordic countries (Mobilitetsstyregruppen, 2010). In fact, the main reasons for moving are employment opportunities and participation in education. Having to move to participate in education may constitute a significant barrier for participation in education for many young Greenlanders and provide an incentive for taking up unskilled work and making it a long-term career. In fact, the main reasons for not moving are social networks of family and friends and employment opportunities at home (Mobilitetsstyregruppen, 2010, p. 12). According to The Economic Council of Greenland (2018, p. 41), there is in fact low geographic labour mobility. Few Greenlanders are willing to move across the country for employment opportunities. It would seem an additional paradox exists. There is a high degree of geographic mobility in general in Greenland but low geographic mobility in terms of moving from areas of high unemployment to areas of low unemployment (The Economic Council of Greenland, 2018, p. 41–43). To increase employment related mobility, it is possible to apply for lump sum payments (mobility enhancing benefits) for several expenses if a worker moves from one part of the country to another for job-related reasons (VIVE, 2019b).

Despite the generally high geographic mobility of Greenlanders, the share of young Greenlanders participating in education is quite low in an international perspective. This is depicted in Figure 9.3.

Figure 9.3 shows the share of 17-to 29-year-olds participating in education in 2018 for the European countries and in 2019 in Greenland. We include only people participating in upper secondary education or further education. The EU28 average amounts to 35.5%. More than one third of the EU28 population between the ages of 17 and 29 years are enrolled in the educational system. Denmark and Finland have a high share of young people in the educational system, roughly 48% in both countries. For Greenland, we depict two bars. The first bar shows the share of 17-to 29-year-old Greenlanders who are studying in Greenland. Of the young Greenlandic population, 18.8% are enrolled in education. The second bar also includes Greenlanders who are studying abroad in addition to those who are studying in Greenland. The share increases to roughly 25%, which is still quite low in comparison with the other countries.

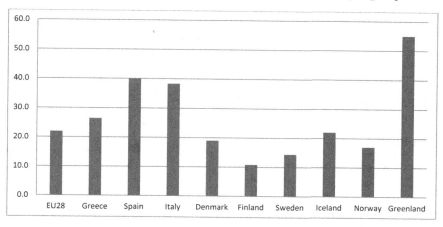

Figure 9.4 Share of the population (25–64 years) with less than primary, primary, and lower secondary education (ISCED levels 0–2).

Source: Statistics Greenland [UDDISCPROH] and Eurostat [edat_lfs_9903].

Historically, the share of Greenlanders participating in education has been even lower, which is reflected in the educational attainment among the 25- to 64-year-olds.

As can be seen from Figure 9.4, the share of the population with low educational attainment in Greenland is much higher than in most other countries. Of the Greenlandic population, 55% of 25- to 64-year-olds have, at most, completed lower secondary schooling, whereas the EU28 average is 22% and 19% in Denmark. A number of publications identify the low educational attainment of the Greenlandic population as a major challenge (e.g., Government of Greenland, 2018; The Economic Council of Greenland, 2019), and a pivotal strategy in Greenlandic labour market policy is improving qualifications and educational attainment of all people who are not in employment, as we elaborate upon later in the chapter. However, the educational attainment of the Greenlandic population is increasing. Over a 10-year period between 2008 and 2018, the share of the population aged 16–74 years with at least lower secondary schooling increased by 6.8 percentage points (Statistics Greenland, 2019a).

9.3 Labour market outcomes for young and older Greenlanders

Having compared the Greenlandic youth employment rate and the NEET rate to selected European countries, the next part of the analysis focuses on comparing the Greenlandic youth (aged 17–29 years) to the older Greenlanders (aged 30–64) on several relevant labour market statistics.

Table 9.1 The Labour Force Survey employment and unemployment rate, educational rate and the NEET rate ("n" in parentheses)

	17–29 years	30–64 years	Total
Employment rate	69.9% (197)	82.1% (505)	78.3% (702)
Not in employment	30.1% (85)	17.9% (110)	21.7% (195)
LFS unemployment rate	3.9% (11)	2.0% (12)	2.6% (23)
In education	16.3% (46)	3.1% (19)	7.3% (65)
Not in employment or education (NEET rate)	16.7% (47)	15.1% (93)	15.6% (140)

Source: The Greenlandic Labour Force Survey.

Table 9.1 compares relevant statistics on labour market outcomes for young and older Greenlanders.

The first row in Table 9.1 outlines the employment rate for Greenlanders aged 17–29 years, Greenlanders aged 30–64 years, and the two groups combined (total).

First, it is worth remembering that the youth employment rate of 69.9% is quite high in an international perspective, as illustrated in Figure 9.2. Among the rest of the working age population (30–64 years), the employment rate is 82.1% – also high in an international perspective. The total employment rate of 78.3% is also quite impressive because the EU28 employment rate was 69.5% in the second quarter of 2019. Conversely, the share not in employment in each of the age groups is 30.1% and 17.9%, respectively.

Before discussing the LFS unemployment rate, we provide a brief explanation of unemployment definitions. The LFS terminology defines unemployed as people without work, who are available for work, and who are actively looking for a job (see Chapter 5 for the full definition). In contrast, we calculate the register-based unemployment rate based on the number of people who are registered at the employment service. In many cases, the two unemployment rates differ markedly (e.g., Statistics Denmark, 2014; Statistics Greenland, 2013). This is also the case in Greenland, where the LFS unemployment rate is lower than the register-based unemployment rate (see, e.g., Government of Greenland, 2020).

Keeping the above in mind, the Greenlandic LFS unemployment rate is quite low in an international perspective: 3.9% among the young, 2.3% among the older population, and 2.6% in total. The LFS unemployment among the EU28 countries amounts to 6.2% in the second quarter of 2019. A low LFS unemployment is, however, not necessarily desirable because we only include people who are unemployed, actively looking for a job, and can start in a new job within two weeks in the measure. Subtracting the LFS unemployed from the share who are not in employment or education, we find that 13% of the working age population are, in a sense, unemployed (not working) and not looking for work.

The share of young Greenlanders who are in education (consider themselves a student) is only 16.3%. This is a low share considering that young people are the primary target group for the educational system (cf., Figure 9.3).

Teenagers from settlements have to move to Nuuk or another one of the aforementioned larger towns to study. As mentioned, limited educational opportunities for further education exist in Greenland. Therefore, 30% of Greenlandic students study abroad, the vast majority of which study in Denmark (Statistics Greenland, 2020a). This causes some "dark" numbers when assessing the educational rate through our survey because the survey included only people residing in Greenland. Increasing the educational rate from Statistics Greenland by 40% yields an educational rate of roughly 25%, which is probably more accurate but still quite low in an international perspective.

Judging solely from the employment rate of young Greenlanders, it would seem that they have worse outcomes than their older working age counterparts do. However, this is not the case when taking account of the educational rate. For young Greenlanders, the share who are in either employment or education is 86.2%, compared to 85.2% for the Greenlanders aged 30–64 years. Taking these two positive outcomes together, young Greenlanders fare just as well as their older counterparts do.

Moving on to the NEETs, the main focus of the chapter, we find that 16.7% of young Greenlanders can be categorized as NEETs (cf. Table 9.1 and Figure 9.1). Of Greenlanders between 30 and 64 years of age, 15.1% are not in education or employment (older NEETs in a sense).

Therefore, we turn our attention to what characterizes the people in Greenland, who are in neither employment or education in comparison to those in employment and education.

Focusing on the NEETs, we see that women are overrepresented in the group because they constitute an estimated 57.5% of the NEETs in Greenland, which indicates a slight over-representation of women in the NEET category. Internationally, women are generally overrepresented in the NEET statistics across the OECD countries (Carcillo et al., 2015 p. 22). Interestingly, we find no differences between genders among the older group outside the labour market and educational system.

Turning to employment among young Greenlanders, we find that women constitute 43.7% of young Greenlanders in work. Thus, young Greenlandic women are in employment to a lesser degree than their male counterparts are. The same is the case for the women aged 30–64 years who constitute 46.7% of those in employment.

Focusing on young Greenlanders who are in education, the Greenlandic LFS estimates 67.4% are women. However, data from Statistics Greenland shows that the share of students who are female amounts to 57.2%. Thus, the LFS overestimates the share of female students. Regardless, young Greenlandic women participate in education to a much larger degree than their male counterparts do.

Table 9.2 Characteristics of NEETs in Greenland ("n" in parentheses)

	17- to 29-year-olds			30- to 64-year-olds		
	Not in employment or education	In employment	In education	Not in employment or education	In employment	In education
Gender						
Women	57.5% (27)	43.7% (86)	67.4% (31)	49.5% (46)	46.7% (236)	89.5% (17)
Men	42.6% (20)	56.4% (111)	32.6% (15)	50.4% (47)	53.3% (269)	10.5% (2)
Total	100% (47)	100% (197)	100% (46)	100% (93)	100% (505)	100% (19)
Educational attainment						
Unskilled (primary and secondary school)	83.0% (39)	66.0% (130)	97.8% (45)	49.5% (46)	26.5% (134)	52.6% (10)
Further education	2.1% (1)	6.1% (12)	2.8% (1)	6.5% (6)	12.1% (61)	15.8% (3)
Skilled worker/ Vocational education	2.1% (1)	16.8% (33)	0.0% (0)	20.4% (19)	42.4% (214)	21.1% (4)
None of the above	12.8% (6)	11.8% (22)	0.0% (0)	23.7% (22)	19.0% (96)	10.5% (2)
Total	100% (47)	100% (197)	100% (46)	100% (93)	100% (505)	100% (19)

Source: The Greenlandic Labour Force Survey.

Lastly, in relation to Table 9.2, we investigate the educational level of NEETs, employed Greenlanders, and Greenlanders participating in education.

From Table 9.2, we see that the vast majority of NEETs (83%) have low educational attainment (having completed primary or secondary school at the maximum). For the older age group also not in employment or education, the share is roughly 50%. The high share of NEETs with no further education is rather unsurprising because NEETs typically have not (yet) completed further education (Mussida and Sciulli, 2018, p. 136; VIVE, 2019a). Furthermore, the youngest in the age group are typically too young to have completed further education or a vocational education. Therefore, it is expected that NEETs have lower educational attainment as a whole, than their older counterparts do who are also not in employment or education.

However, it is quite interesting that half of the older group who are in neither employment nor education have completed some form of education

Table 9.3 Share of NEETs who have applied for a job during the last four weeks

	All groups not in employment or education	NEETs (17–29 years)	Older groups not in employment or education (30–64 years)
Has applied for a job	32.1% (45)	46.8% (22)	24.7% (23)
Has not applied for a job	67.9% (95)	53.2% (25)	75.3% (70)
Total	**100% (140)**	**100% (47)**	**100% (93)**

Source: The Greenlandic Labour Force Survey.

besides mandatory schooling. We further explore some possible reasons for this finding under the subheading of poor health as a barrier for participation.

Next, we investigate how many NEETs are actively looking for a job by showing the share of NEETs who have applied for a job during the course of the last month before the respondent answered the survey. This is relevant because we get an understanding of how many are actually trying to get out of their situations as NEETs.

As can be seen from Table 9.3, roughly a third (32.1%) of all Greenlanders who are not in employment or education have applied for a job during the reference period. However, there are age differences. Almost half (46.8%) of the NEETs (ages 17–29) applied for a job during the reference period. For the older age group, the share is roughly a quarter (24.7%). Thus, job search activity seems higher among the NEETs than their older counterparts not in employment or education do. This is consistent with existing research finding that job search intensity decreases with age (e.g., Banfi et al., 2019; Choi et al., 2015; Menzio et al., 2016). An explanation for this is that, when older workers become unemployed, some stop searching for jobs as a way of early retirement (Choi et al., 2015).

9.3.1 Poor health as a barrier for participation

The Greenlandic LFS also asked a question investigating why the respondents did not apply for a job (not depicted in this chapter. Here, a clear pattern of health-related problems emerged. Taking the categories *Own illness or handicap* and *Is receiving or applying for disability pension* together, almost half (44.7%) of those not in employment or education (young and older), who are not actively searching for a job suffer from health-related issues. This is consistent with existing research showing that NEETs often struggle with health-related issues (Eurofound, 2012); Danish research showing that poor health is a significant barrier for labour market participation among (Danish) cash benefit (social assistance) recipients (Expert Committee, 2015; Jensen, 2014; Ravn, 2019; Ravn & Bredgaard, 2020a); and among socially disadvantaged Greenlanders in Denmark (Statens institut for Folkesundhed, 2019). In fact, socially disadvantaged Greenlanders living in Denmark are twice as likely to experience poverty and homelessness compared to other socially disadvantaged native Danes (Statens Institut for Folkesundhed, 2019).

In relation to health, a vast body of research has found that indigenous peoples have poorer health than nonindigenous people do, and poor health is associated with poverty and unemployment (Durie, 2003; Gracey & King, 2009; King et al., 2009). This is also the case in Greenland, where a study finds that Inuits have poorer health and a lower life expectancy than Danes do (in Denmark; Anderson et al., 2016). In addition, a study of social inequality in health among Inuits in Greenland finds significant variation in health outcomes across Inuit social groups (Bjerregaard et al., 2018). Inuits in a "low" social group (e.g., unemployed and Inuits with low educational attainment) have poorer health outcomes than Inuits in a "high" social group.

Despite many unemployed Greenlanders having health-related problems, data from the labour force survey (not shown in a table) reveals that a small minority of Greenlanders not in education or employment consider themselves long-term ill or disability pensioners. Instead, roughly 85% of young Greenlandic NEETs consider themselves either unemployed or out of work for other reasons. This is the case for roughly 70% of the older group not in employment or education. We find a puzzling discrepancy here. Of those not in employment or education who have not applied for a job, almost half have not done so because of some sort of health-related problem. However, the majority of people not in employment or education mainly consider themselves unemployed. It would thus seem that illness identity (Van Bulck et al., 2018), a term used to describe to which degree an illness dominates a person's self-understanding, is not pronounced among Greenlanders outside the labour market or the educational system. Instead, we can conceptualise a large share of this group as disadvantaged unemployed with work-hindering problems, as low educational attainment and poor health impedes employment prospects.

This is furthermore underpinned by Greenland's public employment service (Majoriaq) categorization of jobseekers in three "match groups" (see Chapter 4). In April 2019, 44% of registered Greenlandic jobseekers were in match group two or three and deemed (temporarily) unable to work due to health-related problems, social problems, or other barriers for labour market participation (Statistics Greenland [ARDLEDMA]). Having a large share of benefit recipients with work-hindering problems is familiar in the majority of Nordic countries. Denmark undertook a series of reforms during the last few decades to improve the work capacity of benefit recipients with work-hindering problems (Expert Committee, 2015; Ravn, 2019; Ravn & Bredgaard, 2020b). However, the country has found no golden policy solution. The share of Danish jobseekers with work-hindering problems who were (temporarily) unable to work in April 2019 amounted to 30%, when including unemployment benefit recipients and cash benefit recipients, and

71%, when focusing only on cash benefit recipients (jobindsats.dk). Judging from these numbers, it would, however, seem that the challenge of unemployed with work-hindering problems is somewhat greater in Greenland than in Denmark. This is furthermore accentuated by the fact that the Greenlandic Labour Market Report of 2016/2017 finds that caseworkers in Majoriaq have a tendency to underestimate the severity the jobseekers' problems, categorizing too many jobseekers as job ready, when they are in reality unable to work for the time being (Government of Greenland, 2018, p. 8).

The Greenlandic Population Survey of 2014 furthermore finds that mental health problems are prevalent among Greenlandic youth. Among the youth, 55.6% of young Greenlandic women and 34.1% of Greenlandic men between 18 and 24 years of age are mentally vulnerable (Dahl-Petersen et al., 2014, p. 67).

Summarizing the findings above, we see that NEETs and the older group not in employment or education are strikingly similar in many ways. Both typically have low educational levels, as is the case with NEETs in general (Carcillo et al., 2015; Eurofund, 2012) and among Danes who are not in employment and receive social assistance (cash benefits) in Denmark (Jensen, 2014; Ravn and Nielsen, 2019). However, there are a larger share of NEETs with, at most, primary and secondary schooling compared to the older group. This is unsurprising because they are not yet old enough to have obtained a degree. Our findings also reveal that a rather large share of the older group not in employment or education have, in fact, completed some sort of further education. The main reason why they are out of work is plausibly health-related barriers (cf., the use of disablement rehabilitation jobs/revalidation [revalidering] in the next section and Chapter 4).

9.4 Greenlandic policy initiatives to reduce the number of NEETS

Prior to outlining the existing policy initiatives in Greenland to reduce the number of NEETs in Greenland, it is worth noting that the Greenlandic employment service implemented a large organizational reform from 1 January 2016 (see Chapter 4). Prior to 2016, counselling centres (Piareersarfiit-centres) handled career counselling and guidance tasks regarding employment and educational opportunities. Furthermore, they operated courses for young people to improve (or obtain) their grades from lower secondary school in order for them to be able to apply for further education. Local labour market offices were responsible for payment of income transferring benefits, job brokering (matching supply and demand), planning of active labour market policies, training courses, and so forth. In many instances, the functions and tasks of the two organisations overlapped, and the government suggested a merge of the two organizations (Government of Greenland, 2014). This became a realty with the creation of Majoriaq-centres, responsible for the prior tasks of Piareersarfiit and the labour market offices. The government implemented the merger in hopes of ensuring a tighter connection between educational and employment services activities (VIVE, 2019b).

The low educational level of Greenlandic NEETs, older benefit recipients, and relatively poor health of clients is a concern among Greenlandic politicians, and this reflected in the labour market policies. Many labour market policies initiated by Majoriaq (the Greenlandic employment service) focus on getting the unemployed to complete lower secondary schooling, developing new skills among unemployed unskilled workers (Copenhagen Economics, 2019; The Government of Greenland, 2018), or improving the work capacity of clients. The Danish Evaluation Institute carried out an evaluation of the Greenlandic training and skills development programme (EVA, 2019a). They find that 60% of those who have completed a Majoriaq course are in employment or education by the next year. Due to the limited number of participants and inadequate registration procedures, it was not possible to estimate the effect of the programme.

A plethora of initiatives exists in Greenland to promote employment or participation in education among the NEET group. We detail a select few of these in more detail in the following.

9.4.1 Policies to promote participation in education or improve job readiness

Most Greenlandic employment policies are classified as occupational training and qualification programmes. As already mentioned, a central policy initiated by the current Majoriaq-centres is to increase the number of young Greenlanders starting (and completing) an education. To achieve this end, there are primary school diploma courses (FA-kurser in Danish) to improve the grades from lower secondary school exams. The target group are often young people who have been in employment for a few years or been inactive after lower secondary school. The courses are typically traditional classroom teaching in the subjects the students need to pass or improve their grades to apply for an education. An evaluation of the courses shows mixed results with a need for more differentiated teaching according to the needs of each student (EVA, 2019a, pp. 77–83).

Majoriaq also offers courses with a focus on obtaining practical work-related skills. These programmes typically involve woodworking, cooking, and personal development activities (EVA, 2019a, p. 84). A larger share of the "not job ready" individuals participate in these types of programmes. The evaluation concludes that there is a need of explicating the purpose and end-goal of the practical courses. However, the practical courses are beneficial in terms of developing personal and social skills. A vast amount of Danish impact evaluations of occupational training and qualification programmes and courses find that there are no employment effects of such programmes for people with no or few work-hindering problems (e.g., Arendt, 2013). Instead, they prolong unemployment by creating "locking-in" effects because participants neglect job searches during participation in these programmes (Arendt, 2013, p. 6). Crucially, these qualification and training programmes are not initiated with the purpose of getting the participants into employment in the short term. They are initiated with the purpose of preparing the participants for education or making them "job ready".

Similar to Denmark and the majority of Nordic countries, Greenland offers students enrolled in full-time education a monthly student grant (income transferring benefit for participating in education), which provides an economic incentive for students to participate in education. The Greenlandic student grant is relatively generous, compared to other Nordic countries (Studiestöd I Norden, 2019), which makes part-time work to support oneself less of a necessity than in many other countries. Greenlandic students are also able to take up favourable student loans with low interest rates (Studiestöd I Norden, 2019).

So-called "special benefits" (særydelser) provide an additional incentive to participate in education abroad. While studying abroad, Greenlandic students are entitled to an annual trip back to Greenland. The government reimburses students for their plane ticket expenses, as well as expenses for necessary accommodation during the journey (Government of Greenland, 2017). As an additional incentive for students to return to Greenland after finishing their studies, the government also pays for expenses for a return trip.

To support Greenlanders living permanently in Denmark, as well as Greenlandic students in Denmark, so-called "Greenlandic Houses" are in the four largest cities in Denmark. The Danish government and the Greenlandic government partly finance the houses. They are multipurpose organizations offering guidance and support for Greenlandic students and Greenlanders living permanently in Denmark (The Greenlandic House in Aalborg, 2020). To carry out these functions, they employ social workers and educational counsellors.

9.4.2 Active labour market policies to promote employment

In relation to active labour market policies in Greenland, there are four major programmes/courses in place: occupational training and qualification (described partly above), counselling and motivation courses, work capacity testing and work capacity assessments, and revalidation (a form of vocational rehabilitation). We will elaborate upon these in the following.

A variant of occupational training and qualification programmes, with a strong focus on promoting employment in the short term, are the PCD-courses [PKU-kurser in Danish] (VIVE, 2019b, pp. 77–79). They have a strong focus on developing skills and qualifications that are useful on the labour market. PCD-courses usually provide the participants with qualifications to carry out a specific job function in different industries (VIVE, 2019b). PCD-courses mostly target the job ready clients in match group one.

Next, we move on to courses for the disadvantaged groups who are (currently) unable to work and struggle with more severe problems and reduced work capacity. For long-term unemployed clients in match group two, people who apply for disability pensions, or disability pensioners having their entitlement for disability pension reassessed, so-called counselling and motivation courses are initiated [VM-kurser in Danish]. The purpose of the counselling and motivation courses is to improve the work capacity [Arbejdsevne in Danish] of participants (VIVE, 2019b, p. 70). The course runs for a duration of three weeks at four hours a day and prepares participants for the initiatives described next.

The next step for the participants in the counselling and motivation courses, or those who have obvious barriers for labour market participation, is to have their work capacity assessed. Work capacity assessments are typically carried out as part of "work capacity testing" [Arbejdsprøvning in Danish] at a public or private workplaces (VIVE, 2019b).

For a minimum of two weeks, the participant carries out unpaid work for an employer to determine the work capacity (and the appropriate number of workers hours) of the client. After the work capacity testing phase, the client attends a meeting with the caseworker and the employer, assessing the work capacity. However, caseworkers find it difficult to convince employers take an unemployed client in work capacity testing (VIVE, 2019b, p. 120).

Upon finishing the work capacity assessment, a form of vocational rehabilitation called revalidation [revalidering] can be initiated, if it is deemed plausible, that the person in question can obtain (some) employment (VIVE, 2018). The Greenlandic revalidation programme is similar to revalidation in Denmark. The Greenlandic Government has an official government target of increasing the number of people participating in revalidation (Government of Greenland, 2014). The purpose of revalidation is either to increase the work capacity of the participant to get the person to obtain work or to retrain workers to different occupations, who cannot stay in their current occupations due to health-related reasons. Revalidation is initiated with a clarification phase, which can include work training courses or educational activities (VIVE, 2018, p. 17). Afterwards, the phase of revalidation begins, which can include training in public or private companies (as a disablement rehabilitation job) or ordinary education in the form of educational revalidation. Participants in revalidation are economically compensated throughout the programme in the form of "revalidation benefit", which is 70% of the wage of an unskilled worker, as stated in the collective agreement (VIVE, 2018, p. 17). If revalidation is carried out as a disablement for rehabilitation jobs in a public or private workplace, participants receive ordinary pay, as stated in the collective agreements, but the state reimburses the workplaces for 80% of their wage expenditures in relation to revalidation (a wage subsidy). Despite the subsidies for employers, caseworkers experience difficulties in relation to convincing, especially private, workplaces to hire a person as part of revalidation (VIVE, 2019b, p. 59).

Revalidation lasts for a maximum of two years but has to be completed as quickly as possible. A regression analysis of revalidation in Greenland based on administrative data shows that, controlling for other factors affecting labour market outcomes, revalidation increases employment among participants. However, we state that the results cannot be interpreted as a causal effect because we did not use a control group (VIVE, 2019b).

9.5 Challenges in relation to completing an education or obtaining work

As outlined above, many policies are in place to encourage young Greenlanders to begin an education nationally and abroad. This is pivotal for the economy because having either a vocational education or a further education increases the likelihood of employment (The Economic Council of Greenland, 2017), as is actually the case in almost every OECD country (OECD, 2016).

Pursuing such a strategy however entails a number of challenges. First, dropout rates are high among Greenlandic students. This is both the case for students studying within the country and among students who go to Denmark to study (EVA, 2019b; Rådgivnings- og Støttecentret, 2019; Statistics Greenland, 2020b, 2020c). Furthermore, 30% of Greenlandic students in further education actually study abroad – the vast majority of which study in Denmark (Statistics Greenland, 2019b). In this regard, it is highly relevant that many Greenlanders studying abroad settle more or less permanently outside of Greenland after obtaining their degrees. More than 45% of those who complete a vocational education or further education abroad live outside of Greenland 10 years after obtaining their degrees (Statistics Greenland, 2019c). In other words, almost half of the students studying abroad settle and make lives for themselves there. This, of course, reduces the total Greenlandic labour supply (see, Chapter 2). The total number of people residing in Greenland amounted to roughly 56,000 in 2018, but there were nearly 16,500 people born in Greenland living in Denmark in the same year (Statistics Greenland, 2020a). Almost 30% of the people born in Greenland live outside of Greenland, which poses a major challenge for Greenland in a situation with strong demand for labour.

The number of Greenlanders studying abroad has increased by 21% since 2001, while the share of students returning to Greenland has remained constant. Increasing the absolute number of students who complete an education (domestically and abroad) is thus still a feasible strategy to pursue.

This is accentuated by the fact that unemployment in Greenland is almost non-existent among people who have completed further education (The Economic Council of Greenland, 2017). However, the Greenlandic politicians have a clear incentive for increasing the share who study (and complete an education) domestically. This is underpinned by data from Statistics Greenland showing that nearly 90% of Greenlanders who complete an education domestically still live in Greenland 10 years after obtaining their degrees (Statistics Greenland, 2019c). This, of course, provides an incentive for the Greenlandic government to encourage young people to study domestically. As already mentioned, Greenland has rather limited educational opportunities, which makes it necessary to have some students study abroad to maintain a supply of the labour that is in demand.

Turing to the policies that focus on getting clients into employment, a series of challenges is also evident. As stated in relation to work testing and revalidation, many caseworkers find it challenging to engage employers and convince them to take in a client for work testing or revalidation.

Furthermore, Majoriaq caseworkers and managers often experience that there are only a limited number of courses they can offer, and the available courses do not always match the needs of the clients (VIVE, 2019b, p. 80). Caseworkers and managers also find that they are often understaffed and that sickness absence is a major problem among Majoriaq staff (VIVE, 2019b, pp. 82–83). Due to the shortage of staff, clients who are in need of a course or programme are often not offered to participate in one or must wait a long time to participate (VIVE, 2019b, p. 74).

From an active labour market research perspective, it is striking that the absence of work place training programmes and wage subsidy programmes (except for revalidation as disablement rehabilitation jobs) characterise Greenlandic labour market policies because these have been found effective in bringing job ready and disadvantaged unemployed into employment, especially, if they take place in the private sector (e.g., Graversen, 2012; Rosholm & Svarer, 2011). However, it is questionable if these programmes work for the clients with severe problems (Skipper, 2010). In Denmark, workplace training programmes take the form of unpaid internships, where the unemployed work for an employer for a number of weeks (usually 6–13 weeks) while receiving public income transferring benefits. In the Danish wage subsidy programme, an employer hires an unemployed person on ordinary terms but is reimbursed for some of the wage expenditures. In both programmes, the employers get to experience and assess the work effort and competencies of the participant in practice before making a final decision about hiring the person in unsupported employment. However, considering the apparent difficulties of engaging Greenlandic employers in existing active labour market polices, it is uncertain if these types of programmes are feasible to pursue.

9.6 Conclusions

In this chapter, we explored NEETs in Greenland from a national and international perspective. Overall, we find that Greenland has a higher NEET rate than the EU average, and a much higher NEET rate than the other Nordic countries. However, Greenlandic youth employment is higher than most other European countries, whereas participation in education is lower. The scarce educational opportunities in Greenland, motivating Greenlanders to study abroad – primarily in Denmark, can explain, to some extent the lower educational rate.

We furthermore compared Greenlandic NEETs (17–29 years) and older (30–64 years) disadvantaged groups not in employment or education. We found that NEETs and their older counterparts are quite alike in many respects. However, a larger share of NEETs are actively looking for work compared to the older age group. The main reason for not applying for a job is poor health, which is consistent with existing research finding poor health among a large share of the Greenlandic population as a major issue.

Compared to the population in employment, NEETS and the older group not in employment or education have lower educational attainment (cf., Table 9.2). This is unsurprising because people with low educational attainment are overrepresented in unemployment statistics internationally (OECD, 2019). Greenlandic NEETs and the older group not in employment or education largely consist of disadvantaged people with poor health and low educational attainment, which impedes employment.

Similar to the vast majority of OECD countries, the Greenlandic government has a strong focus on reducing the NEET group, and several policies are in place to encourage young people to participate in education or obtain employment. A lack of programmes matching the needs of clients, shortages of staff at employment services, sickness absence among staff, and a long waiting list for programmes are among the challenges at hand. The employment system has undergone a rather recent organizational reform, and time will tell if it will improve employment outcomes for NEETs and other disadvantaged groups.

There are no clear-cut policy solutions for increasing employment, labour supply, or educational attainment in Greenland, but strengthening and developing new policies to combat youth unemployment and to increase participation in education and to reduce dropouts could be a strategy to pursue.

References

Anderson, I. et al. (2016). Indigenous and tribal peoples' health (The Lancet–Lowitja Institute Global Collaboration): A population study. *The Lancet, 388*(10040), 131–157.

Arendt, J. N. (2013). *Effekter af kurser med vejledning og særligt tilrettelagt opkvalificering for ledige – En oversigt over danske og internationale kvantitative studier* [Effects of courses with guidance and specially designed upskilling for the unemployed - An overview of Danish and international quantitative studies]. Copenhagen: KORA.

Arnfjord, S. (2016). Sociale forhold og socialpolitik i Grønland [Social conditions and social policy in Greenland]. In I. H. Møller & J. E. Larsen (Eds.), *Socialpolitik*. Copenhagen: Hans Reitzels Forlag.

Banfi, S., Choi, S., & Villena-Roldán, B. (2019). Deconstructing job search behavior. *University of Bristol Discussion Paper, 19*, 706.

Bjerregaard, P., Dahl-Petersen, I. K., & Larsen, C. V. L. (2018). Measuring social inequality in health amongst indigenous peoples in the Arctic. A comparison of different indicators of social disparity among the Inuit in Greenland. *SSM – Population Health, 6*, 149–157.

Carcillo, S., Fernández, R., & Königs, S. (2015). NEET Youth in the Aftermath of the Crisis: Challenges and Policies. *OECD Social, Employment and Migration Working Papers* No. 164. OECD.

Choi, S., Janiak, A., & Villena-Roldán, B. (2015). Unemployment, participation and worker flows over the life-cycle. *The Economic Journal, 125*(589), 1705–1733.

Copenhagen Economics. (2019). *Evaluering af beskæftigelsesstrategi 2015*. Greenland: Naalakkersuisut.

Dahl, J. (2010). Identity, urbanization and political demography in Greenland. *Acta Borealia, 27*(2), 125–140. DOI: 10.1080/08003831.2010.527528.

Dahl-Petersen, I. K., Viskum, C., Larsen, L., Nielsen, N. O., Jørgensen, M. E., & Bjerregaard, P. (2014). *Befolkningsundersøgelsen i Grønland 2014. Levevilkår, livsstil og helbred* [The population survey in Greenland 2014. Living conditions, lifestyle and health]. Copenhagen: Statens Institut for Folkesundhed, Syddansk Universitet.

Durie, M. H. (2003). The health of indigenous peoples. *BMJ, 326*(7388), 510–511. https://doi.org/10.1136/bmj.326.7388.510

Eurofound. (2012). *NEETs – Young people not in employment, education or training: Characteristics, costs and policy responses in Europe.* Publications Office of the European Union, Luxembourg.

Eurofound. (2017). *Long-term unemployed youth: Characteristics and policy responses.* Publications Office of the European Union, Luxembourg.

EVA. (2019a). *Evaluering af vejlednings- og opkvalificeringsindsatser i Majoriaq* [Evaluation of guidance and retraining efforts in Majoriaq]. The Danish Evaluation Institute, Copenhagen.

EVA. (2019b). *Grønlandske studerende på videregående uddannelser i Danmark* [Greenlandic students in higher education in Denmark]. The Danish Evaluation Institute, Copenhagen.

Expert Committee. (2015). *Nye veje mod job – for borgere i udkanten af arbejdsmarkedet* [New paths to jobs - for citizens on the fringes of the labour market]. Ekspertgruppen om udredning af den aktive beskæftigelses indsats.

Gallie, D., Paugam, S., & Jacobs, S. (2003). Unemployment, poverty and social isolation. Is there a vicious circle of social exclusion? *European Societies, 5*(1), 1–32.

Government of Greenland. (2014). *One Country – One labour market – Employment strategy 2014–2017.* https://naalakkersuisut.gl/~/media/Nanoq/Files/Hearings/2014/Beskaeftigelsesstrategi/Documents/Et%20land%20et%20arbejdsmarked%20Besk%c3%a6ftigelsesstrategi%202014-2017_050814_DK.pdf

Government of Greenland. (2015). *A safe labour market. Employment strategy 2015.* https://naalakkersuisut.gl/~/media/Nanoq/Files/Publications/Arbejdsmarked/DK/Beskaeftigelsesstrategi_DK_01102015.pdf

Government of Greenland. (2017). *Selvstyrets bekendtgørelse nr. 9 af 19. juli 2017 om uddannelsesstøtte.* https://www.sumut.dk/media/2318/selvstyrets-bekendtg%C3%A3-relse-nr-9-af-19-juli-2017-om-uddannelsesst%C3%A3-tte.pdf

Government of Greenland. (2018). *Labour Market Report 2016–2017.* https:// naalakkersuisut.gl/~/media/Nanoq/Files/Publications/Arbejdsmarked/DK/ Arbejdsmarkedsredeg%C3%B8relse%202016-17%20DK.pdf

Government of Greenland. (2020). *Labour Market Report 2018–2019.* https:// naalakkersuisut.gl/~/media/Nanoq/Files/Publications/Arbejdsmarked/DK/ Arbejdsmarkedsredeg%c3%b8relsen%202018-2019%20DK.pdf

Gracey, M., & King, M. (2009). Indigenous health part 1: Determinants and disease patterns. *The Lancet, 374*(9683), 65–75.

Graversen, B. K. (2012). *Effekter af virksomhedsrettet aktivering for udsatte ledige.* SFI.

Greenlandic Ministry of Finance. (2019). *Political Economic Report, Ministry of Finance.* https://naalakkersuisut.gl/~/media/Nanoq/Files/Attached%20Files/Finans/ DK/Politisk%20Oekonomisk%20Beretning/P%C3%98B2019%20DK.pdf

Heldring, L., & Robinson, J. A. (2012). Colonialism and Economic Development in Africa. *NBER Working Paper No. 18566.*

Henderson, J. L., Hawke, L. D., & Chaim, G. (2016). Not in employment, education or training: Mental health, substance use, and disengagement in multi-sectoral sample of service-seeking Canadian youth. *Children and Youth Services Review, 75,* 138–145.

Jensen, B. (2014). *Hvad ved vi om modtagerne af kontanthjælp?* [What do we know about the recipients of cash benefits]. Copenhagen: Syddansk Universitetsforlag.

King, M., Smith, A., & Gracey, M. (2009). Indigenous health part 2: The underlying causes of the health gap. *The Lancet, 374*(9683), 76–85.

Maguire, S. (2015). Young people not in education, employment or training (NEET): Recent policy initiatives in England and their effects. *Research in Comparative & International Education, 10*(4), 525–536.

Maguire, S., & Thompson, J. (2007). Young people not in education, employment or training (NEET) – Where is government policy taking us now? *Youth & Policy, 8*(3), 5–18.

Menzio, G., Telyukova, I. A., & Visschers, L. (2016). Directed search over the life cycle. *Review of Economic Dynamics, 19*(Special Issue), 38–62.

Mobilitetsstyregruppen. (2010). Mobilitet I Grønland. *Sammenfatning af hovedpunkter fra analysen af mobilitet i Grønland* [Mobility In Greenland. Summary of main points from the analysis of mobility in Greenland]. Nuuk: Mobilitetsstyregruppen.

Mussida, C., & Sciulli, D. (2018). Labour market transitions in Italy: The case of the NEET. In M. A. Malo & A. M. Mínguez (Eds.), *European Youth Labour Markets.* New York: Springer International Publishing.

Paul, K. I., & Moser, K. (2009). Unemployment impairs mental health: Meta-analyses. *Journal of Vocational Behavior, 74*(3), 262–282.

OECD. (2015). *Employment Outlook 2015.* OECD Publishing, Paris. http://dx.doi. org/10.1787/empl_outlook-2015-en

OECD. (2016). The NEET challenge: What can be done for jobless and disengaged youth? In *Society at a Glance 2016* (pp. 13–68). Paris: OECD.

OECD. (2019). *Education at a glance 2019: OECD indicators.* Paris: OECD Publishing. https://doi.org/10.1787/f8d7880d-en

Ravn, R. (2019). *Beskæftigelsesrettet rehabilitering: En evaluering af Hjørring Kommunes investering på beskæftigelsesområdet* [PhD dissertation]. Aalborg University, Denmark.

Ravn, R., & Nielsen, K. (2019). Employment effects of investments in public employment services for disadvantaged social assistance recipients. *European Journal of Social Security*, *21*(1), 42–62.

Ravn, R. L., & Bredgaard, T. (2020a). Relationships matter—The impact of working alliances in employment services. *Social Policy and Society*. https://doi.org/10.1017/S1474746420000470

Ravn, R. L., & Bredgaard, T. (2020b). Beskæftigelsesrettet rehabilitering: En ny tilgang til beskæftigelsesindsatsen for udsatte personer uden for arbejdsmarkedet? [Employment-oriented rehabilitation: A new approach to employment efforts for vulnerable people outside the labor market]. In M. P. Klindt, S. Rasmussen, & H. Jørgensen (Eds.), *Aktiv arbejdsmarkedspolitik: Etablering, udvikling og fremtid* (pp. 301–320). Djøf Forlag. Arbejdsmarkedsforhold No. 2.

Rosholm, M., & Svarer, M. (2011). *Effekter af virksomhedsrettet aktivering i den aktive arbejdsmarkedspolitik [Effects of business-oriented activation in it active labor market policies] in SFI (2015) Ressourceforløb-Koordinerende sagsbehandleres og borgeres erfaringer*. København: Arbejdsmarkedsstyrelsen [The Danish Labour Market Agency].

Rud, S. (2019). Mod Bedre Vidende – Grønland og Politisk-Ideologisk Historieskrivning. *Temp - Tidsskrift for Historie*, *9*(17), 145–152.

Rådgivnings- og støttecentret. (2019). *Udvikling, gennemførsel og evaluering af en studiestøttende indsats for studerende med grønlandsk sproglig og kulturel baggrund i videregående uddannelse i Region Midt* [Development, implementation and evaluation of a study support effort for students with a Greenlandic linguistic and cultural background in higher education in the Central Region]. Aarhus: Aarhus University.

Skipper, L. (2010). *En mikroøkonometrisk evaluering af den aktive beskæftigelsesindsats* [A microeconometric evaluation of active employment efforts]. AKF, Copenhagen.

Staiger, T., Waldmann, T., Oexle, N., Wigand, M., & Rüsch, N. (2018). Intersections of discrimination due to unemployment and mental health problems: The role of double stigma for job- and help-seeking behaviors. *Social Psychiatry and Psychiatric Epidemiology*, *53*, 1091–1098.

Statens Institut for Folkesundhed. (2019). *Sundhedsprofil for socialt udsatte grønlændere i Danmark*. *Statens Institut for Folkesundhed* [Health profile for socially disadvantaged Greenlanders in Denmark. The National Institute of Public Health]. Denmark: Syddansk Universitet.

Statistics Denmark. (2014). *Danmarks Statistiks forskellige ledighedsbegreber*. https://www.dst.dk/ext/arbe/ledighedsbegreber–pdf

Statistics Greenland. (2013). *Bemærkninger til arbejdsmarkedsstatistikkerne*. https://stat.gl/publ/da/AR/201303/pdf/Bem%C3%A6rkninger%20til%20arbejdsmarkedsstatistikkerne.pdf

Statistics Greenland. (2019a). *The educational profile of the population 2018*. http://www.stat.gl/publ/da/UD/201907/pdf/Befolkningens%20uddannelsesprofil%202018.pdf

Statistics Greenland. (2019b). *Further education 2018*. http://www.stat.gl/publ/da/UD/201903/pdf/2018%20videregående%20uddannelser.pdf

Statistics Greenland. (2019c). *Country of residence after completion of education*. http://www.stat.gl/publ/da/UD/201908/pdf/2018%20Bopæl%20efter%20endt%20uddannelse.pdf

Statistics Greenland. (2020a). *Greenland in numbers.* https://stat.gl/publ/da/GF/2020/pdf/Gr%C3%B8nland%20i%20tal%202020.pdf

Statistics Greenland. (2020b). *Completion of secondary education.* http://www.stat.gl/dialog/main.asp?lang=da&sc=UD&version=202004

Statistics Greenland. (2020c). *Completion of further education.* http://www.stat.gl/dialog/main.asp?lang=da&sc=UD&version=202005

Studiestöd i Norden. (2019). *Studerende i Norden - studiestøtte og økonomi.* http://www.studiestodinorden.org/wp-content/uploads/2019/01/ASIN-rapport-Studerende-i-Norden-studiest%C3%B8tte-og-%C3%B8konomi.pdf

The Economic Council of Greenland. (2017). *The Greenlandic Economy 2017.*

The Economic Council of Greenland. (2018). *The Greenlandic Economy 2018.*

The Economic Council of Greenland. (2019). *The Greenlandic Economy 2019.*

The Greenlandic House in Aalborg. (2020). *Det Grønlandske Hus i Aalborg.* https://dgh-aalborg.dk/in-english-1

Thisted, K. (2016). The specter of Danish empire: The prophets of eternal fjord and the writing of Danish-Greenlandic history. In J. Björklund & U. Lindqvist (Eds.), *New Dimensions of Diversity in Nordic Culture and Society* (pp. 151–173). Cambridge: Cambridge Scholars Publishing.

Van Bulck, L., Luyckx, K., Goossens, E., Oris, L., & Moons, P. (2018). Illness identity: Capturing the influence of illness on the person's sense of self. *European Journal of Cardiovascular Nursing, 18*(1), 4–6.

VIVE. (2018). *Analyse af offentlig hjælp. Det nationale forsknings- og analysecenter for velfærd* VIVE.

VIVE. (2019a). *Unge uden job og uddannelse – hvor mange, hvorfra, hvorhen og hvorfor? En kortlægning af de udsatte unge i NEET-gruppen.* Det nationale forsknings- og analysecenter for velfærd. VIVE.

VIVE. (2019b). *Evaluering af Majoriaq-centrenes arbejdsmarkedsindsats.* Det nationale forsknings- og analysecenter for velfærd. VIVE.

Yates, S., Harris, A., Sabates, R., & Staff, J. (2011). Early occupational aspirations and fractured transitions: A study of entry into 'NEET' status in the UK. *Journal of Social Policy, 40*(3), 513–534.

Index

Note: *Italicized* and bold pages refer to figures and tables, respectively.

Printed in the United States
by Baker & Taylor Publisher Services